Introduction to Human Gross Anatomy

A Regional Laboratory Guide

Second Edition

Dr. Mark D. Womble
Department of Biological Sciences
Youngstown State University

Kendall Hunt
publishing company

www.kendallhunt.com
Send all inquiries to:
4050 Westmark Drive
Dubuque, IA 52004-1840

Previously entitled *A Laboratory Guide for Introduction to Human Gross Anatomy*

ISBN 978-0-7575-7921-9

Printed in the United States of America
10 9 8 7 6 5 4 3

CONTENTS

INTRODUCTION – COMPUTER SOFTWARE AND LABORATORY GUIDES

Introduction to Human Gross Anatomy is a lecture and laboratory course that provides a detailed overview to the structure of the human body. The laboratory is self-paced and is based on two computer software programs, ***AnatLab: The Anatomy Lab*** and ***ADAM Interactive Anatomy***. The software programs are available on the computers found in the Computer Teaching Laboratory, or they may be purchased through the YSU Bookstore.

This Laboratory Guide follows a regional approach to the study of human anatomy and is divided into four major sections: The Lower Limb, The Head and Neck, The Back and Upper Limb, and The Abdomen, Pelvis, and Thorax. For each lab throughout the course, always do the ***AnatLab*** software chapter first. This is a very linear software that gives a thorough, introductory overview of the material for each lab guide chapter. Next, follow the instructions provided in this Laboratory Guide to identify structures, using the ***ADAM Interactive Anatomy*** software. Use *ADAM* to help you develop a three-dimensional understanding of the positional relationships for each structure within the body.

As you work through the software lab exercises, have your lecture notes available so that you can read about each structure as you identify it. Use the bones, skeleton, and anatomical models to review muscle attachment points, muscle actions, innervations, blood supplies, and positional relationships. A full, three-dimensional understanding of body structure cannot be accomplished by only examining flat, two-dimensional pictures. In addition to using the skeleton and models, think about where each structure is located within your own body. Remember, you are a walking cheat-sheet for this course!

ANATLAB

AnatLab (Elephant CDs: 614-595-1591) is an interactive, multimedia, tutorial software program designed to help the user learn the fundamentals of human anatomy. It provides a guide for all major anatomical structures and will help you to gain a three-dimensional understanding of the human body. However, it does not cover every anatomical structure and thus is not an atlas of the entire body.

The program is divided into four major body regions, with several chapters of material for each region. It includes simple illustrations, still images, radiographs (X-ray pictures), and video clip demonstrations of real cadaver material. Verbal pronunciation of many anatomical terms is also provided. Embedded questions and a short quiz at the end of each chapter serve to reinforce ther material to be learned. The program is designed to be used in a linear fashion. When you have completed a lesson, you will have a good basic understanding of the major anatomical structures in that region.

Features:

1. **Highlighted words** can be clicked on within the text to hear pronunciations.
2. The **Continue** button proceeds to the next screen page. The **Reverse** button goes back one page.
3. Questions appear frequently within each chapter. Type in your response, and press **Enter** to view the answer. A response must be provided before moving to the next page.
4. The **Help** button will remind you of these features and allow you to return to your position in the program.
5. When a video clip movie is available, the **Demonstration** button will appear in the text box. You can click on it to see the movie as many times as you would like.
6. The **Outline** button takes you to the outline of the lesson. A yellow box surrounds your current topic location. Click on any item to go directly to that subject. Asterisks mark the position on or near a Demonstration movie.
7. A short quiz is given at the end of each chapter. You must answer the questions to continue. At the end of the quiz, feedback will tell you the percentage of correct responses and the items you missed. A **Restart** button will then allow you to go to the outline to find and review those subjects you missed.

ADAM INTERACTIVE ANATOMY

ADAM Interactive Anatomy (adameducation.com) is a highly detailed anatomical software program that offers tens of thousands of colorful anatomical drawings, 3D images, cadaver photographs, and pinned atlas images. *ADAM Interactive* provides a non-linear approach to learning, allowing you to move easily between hundreds of layers of structures in four different views. Each anatomical structure is identified with a simple click of the mouse button. The majority of each Laboratory Guide concentrates on guiding you through *ADAM Interactive*, with detailed descriptions of precisely where to go and what to identify.

Features:

1. There are three components to *ADAM Interactive Anatomy* that will be used:
   a) *Dissectible Anatomy*: This provides an anterior, posterior, medial, or lateral view of either the male or female body. These views can be dissected through multiple layers, one layer at a time. There are hundreds of layers in each view.
   b) *Atlas Anatomy*: This provides a single anatomy atlas type figure. Each structure in the image is individually pinned.
   c) *3D Anatomy*: This features freely rotatable, three-dimensional views of selected body structures or systems.

Using ADAM Interactive Anatomy:

1. Open *ADAM* and choose one of the three components (Dissectible Anatomy, Atlas Anatomy, or 3D Anatomy). This will take you to the **Open** menu, where you can also select among these components by clicking the appropriate tabs.
2. *Dissectible Anatomy*
   a) Select **Male** or **Female** body and **Anterior**, **Posterior**, **Lateral**, **Medial**, **Lateral Arm**, or **Medial Arm** view. Click **Open**.
   b) Use the **Maximize** button (*middle button at top left*) to enlarge the view.
   c) Use the **Depth bar** (located at left of image) to move to different depth layers. Increase or decrease depth by clicking the arrows at the top or bottom of the depth bar, or by sliding the depth bar button up or down. The **Depth layer** is given as a number on the depth bar button.
   d) There are two ways to identify structures:
      1) Place the mouse cursor on a structure and click the left mouse button.
      2) Find the name of a structure in the pull-down menu located just above the image and click on the structure name. The image will automatically go to an appropriate depth layer, and the structure will be highlighted in color (the rest of the image will go to grays). Return to the normal viewing mode by clicking the **Normal mode** button (located in button group to left of image).
   e) With the **Identify** button depressed, structures can be identified by simply placing the mouse cursor on the structure and clicking the left mouse button.
   f) Use the **View** button to change the image view direction (anterior, posterior, medial, or lateral).
   g) Switch between male and female using the **Gender** button.
   h) Increase or decrease magnification of the image with the **Zoom** button. Click this button (cursor changes to a (+) or (–) magnifying glass); then click the cursor on the image.
3. *Atlas Anatomy*
   Select atlas figures by Region or System. Each figure has numerous, individually pinned structures. Identify structures by clicking on the pin or by using the pull-down menu at the top of the image.
4. *3D Anatomy*
   Some body structures or systems are available for three-dimensional viewing. These may be rotated using the arrow buttons to the left of the image. Individual structures must be identified using the pull-down menu at the top of the image.

LABORATORY 1

Introduction to Anatomy; Bones of the Lower Limb

Section I: ***AnatLab*** – Do the chapter for Bones of Lower Limb
Section II: ***ADAM Interactive*** **– Bones of the Lower Limb**

1. Pull down **File** menu – **Open** – **Atlas Anatomy** – **Region** = **Lower Limb** – **Body System** = **Skeletal**. Click the **Details** button, and select **Bones of the Lower Limb (Ant)**. Identify:
   a) Body of femur
   b) Body of ilium
   c) Crest of ilium
   d) Head of femur
   e) Head of fibula
   f) Ischium
   g) Patella
   h) Pubic tubercle
   i) Sacrum
   j) Tuberosity of tibia
2. Close window when finished.
3. **File** menu – **Open** – **Atlas Anatomy** – **Region** – **Lower Limb**. Select **Bones of the Lower Limb (Post)**. Identify the following:
   a) Body of ilium
   b) Body of ischium
   c) Body of pubis
   d) Coccyx
   e) Femur
   f) Head of fibula
   g) Sacrum
   h) Medial and lateral condyles of tibia

ATLAS ANATOMY – COMPARISON OF MALE AND FEMALE PELVIS

Anterior Views of the Pelvis

1. Pull down **File** menu. Select **Open**, then **Content**. Click tab for **Atlas Anatomy**.
2. Select images for **Body Region** = **Pelvis and Perineum**, **Body System** = **Skeletal**. Scroll down and open **Female Bony Pelvis (Ant)**.
3. Click **Zoom** button, and then click on view of pelvis to reduce size of image.
4. Open similar view of **Male Bony Pelvis (Ant)**.

a) **File – Open – Content – Atlas Anatomy – System – Skeletal System – Male Bony Pelvis (Ant)**
b) **Zoom** button – reduce size of male pelvis image

5. Adjust positions so that both images can be seen at same time. Pull down **Window** menu – select **Tile.**
6. On both the male and female pelvis, identify the following landmarks. (Make sure that the **Identify**

button is depressed.)

a) Sacrum
b) Hipbone – *not labeled*; consists of
   1) Pubis – Body, Superior ramus, Pubic tubercle
   2) Ischium – Ischial tuberosity
   3) Ilium – Iliac fossa, Iliac crest, Anterior superior iliac spine, Anterior inferior iliac spine
c) Sacroiliac joint
d) Obturator foramen
e) Acetabulum
f) Ischiopubic ramus (*not labeled*; formed by union of Ischial ramus and Inferior pubic ramus); note difference in the ***subpubic angle*** formed by right-left Ischiopubic rami as they come together at the midline. Which pelvis has an angle of approximately 60°? Which is approximately 90°?

7. When finished, close both pelvis windows.

Lateral Views of the Pelvis

1. Follow the procedure listed in Steps 1-5 above to open views of the **Male-Female Bony Pelvis (Lat)**. Identify the following:
   a) Sacrum and Coccyx
   b) Pubis – Pubic tubercle, Superior pubic ramus
   c) Ischium – Ischial tuberosity, Ischial spine
   d) Ilium – Iliac crest, Anterior superior iliac spine, Anterior inferior iliac spine, Posterior superior iliac spine, and Posterior inferior iliac spine
   e) Greater sciatic notch, Lesser sciatic notch; note that the Greater sciatic notch is broader in females
   f) Acetabulum
2. When finished, close both pelvis windows.

Superior Views of the Pelvis

1. Follow the procedure listed in Steps 1-5 above to open views of the **Male and Female Bony Pelvis (Sup)**. Identify the following:
   a) Ilium – Iliac crest, Anterior superior iliac spine, Anterior inferior iliac spine, Posterior superior iliac spine
   b) Pubis – Pubic tubercle, Superior pubic ramus, Pubic symphysis
   c) Ischium – Ischial spine
   d) Sacrum and Coccyx
   e) Sacroiliac joint
   f) Lesser (below pelvic brim) and Greater (above pelvic brim) pelvic cavities (*not labeled*)
   g) Pelvic brim (*not labeled*) – bony margin that separates lesser and greater pelvic cavities
   h) Note differences in size of male/female Inferior pelvic outlet (inferior opening). (Importance?)
2. When finished, close both pelvis windows.

LABORATORY 2

Muscles of the Hip

Section I: *AnatLab* – Do the chapter for Muscles of the Hip

Section II: *ADAM Interactive* – Bones of the Hip

Anterior View of the Lower Limb

1. Open Dissectible Anatomy window:
   a) Pull down **File** menu – select **Open** – select **Content**
   b) Click tab for **Dissectible Anatomy**
   c) Select **Anterior** view and **Open**
2. Center image over pelvic region.
3. Use **Depth Bar** – go to **layer 180** (if image disappears or becomes black/white, press **Normal Mode** button). Identify the following:
   a) Iliac crest
   b) Pubis
   c) Pubic symphysis
4. Continue to remove layers, stopping at **layers 208** and **266** to again identify the Iliac crest, Pubis, and Pubic symphysis.
5. Go to **layer 300** and identify the Ventral sacroiliac ligament.
6. Go to **layer 329** and identify the following:
   a) Body and transverse processes of L5 vertebra
   b) Sacrum and Coccyx
   c) Parts of hipbone:
      1) Ilium – Iliac crest, Anterior superior iliac spine, Anterior inferior iliac spine
      2) Pubis and Pubic symphysis
      3) Ischium
      4) Obturator foramen (covered by Obturator membrane)
      5) Acetabulum
   d) Sacrospinous ligament and Sacrotuberous ligament
   e) Femur – Head, Neck, Greater trochanter, Lesser trochanter
7. Go to **layer 330** and identify the following:
   a) Ilium – Body, Iliac crest, Iliac fossa
   b) Pubis – Pubic tubercle, Superior ramus
   c) Articular cartilage of hip joint (lining Acetabulum and covering Head of Femur)
   d) Head and Neck of Femur

Posterior View of the Lower Limb

1. Click **View** button and change orientation to **Posterior**.
   a) Click button for **Normal Mode**
   b) Reposition screen to show Posterior hip and pelvic regions
2. Change depth to **layer 117**. Identify the following:
   a) Ilium – Iliac crest, Posterior superior iliac spine
   b) Ischium – Ischial tuberosity
   c) Pubis – Body
   d) Sacrum and Coccyx
   e) Sacrotuberous ligament
3. Change depth to **layer 176**. Identify the following:
   a) Ilium – Posterior inferior iliac spine
   b) Ischium – Ischial spine
   c) Pubis – Superior pubic ramus
   d) Ischiopubic ramus (*not labeled*) – formed by union of Ischial ramus and Inferior pubic ramus
   e) Greater sciatic notch, Lesser sciatic notch
   f) Sacrospinous ligament
   g) Obturator foramen (covered by Obturator membrane)
   h) Sacrum and Coccyx
4. Change depth to **layer 185**. Identify the following in the Femur:
   a) Head and Neck
   b) Greater trochanter & Lesser trochanter
   c) Intertrochanteric crest
   d) Gluteal tuberosity
5. Change depth to **layer 186**. Identify the Head of Femur within the Acetabulum of Hipbone.

RADIOGRAPHIC VIEWS OF THE LOWER LIMB

Atlas Anatomy

1. Pull down **File** menu.
   a) **Open – Content – Atlas Anatomy – System – Skeletal**
   b) Scroll down and open **Pelvis (Ant/Post)**
   c) Use **Zoom** button to reduce size of image
2. Identify the following:
   a) Pubis – Superior pubic ramus, Pubic symphysis
   b) Ischium – Ischial tuberosity
   c) Ilium – Iliac crest
   d) Sacroiliac joint
   e) Greater sciatic notch
   f) Acetabulum
   g) Head and Neck of femur
   h) Greater trochanter and Lesser trochanter of femur
3. Close pelvis window when finished.

Section III: *ADAM Interactive* – Muscles of the Hip

Posterior View of Hip Muscles

1. Open a Posterior view of hip region in Dissectible Anatomy.
   a) **File – Open – Content – Dissectible Anatomy – Posterior**
   b) Position image to posterior pelvic region
2. Start with **Depth Bar** at **layer 0**. Remove the skin, subcutaneous fascia and veins, and cutaneous nerves, one layer at a time. At **layer 7**, identify Fascia Lata of the Thigh, Gluteal Fascia (*not labeled*; included here as part of Fascia lata), and the Crural Fascia of the Leg.
3. At **layer 8**, identify Gluteus Maximus muscle, Iliotibial Tract, and fascia covering Gluteus medius muscle.
4. Change the depth to **layer 104**; identify Gluteus Medius muscle.
5. Change to **layer 105**. Identify the following:
   a) Gluteus minimus muscle
   b) Piriformis muscle (with Sciatic nerve emerging at inferior border of Piriformis)
   c) Superior gemellus and Inferior gemellus muscles
   d) Quadratus femoris muscle
   e) Sacrotuberous ligament
6. Change depth to **layer 121**. Identify the following:
   a) Superior and Inferior gemellus muscles
   b) Obturator internus muscle (from inner surface of Obturator foramen/membrane – exits Pelvis via Lesser sciatic foramen)
   c) Quadratus femoris muscle
   d) Sacrospinous ligament
7. Change depth to **layer 125**. Identify Obturator Externus muscle (coming from outer side of Obturator foramen/membrane).

Lateral View of Hip Muscles

1. Use the **View** button to change to a Lateral view of the hip region. Click **Normal Mode** button.
2. Use the **Depth Bar** to remove skin and subcutaneous structures. Identify Fascia lata.
3. Change depth to **layer 11**. Identify the following:
   a) Gluteus maximus muscle
   b) Tensor fasciae Latae muscle
   c) Iliotibial tract (What is the inferior attachment point for the Iliotibial tract?)
   d) Fascia over Gluteus medius muscle
4. Change depth to **layer 91**. Identify the following:
   a) Gluteus medius
   b) Gluteus minimus
   c) Greater trochanter of femur
   d) Sacrotuberous ligament

5. Change depth to **layer 131**. Identify the following:
   a) Piriformis muscle
   b) Sacrospinous ligament
   c) Greater sciatic notch (Borders of Greater sciatic foramen? Difference from Greater sciatic notch?)

Anterior View of Hip Muscles

1. Use the **View** button to change to an Anterior view of the hip region. Click **Normal Mode** button.
2. Change depth to **layer 266**. Identify the following:
   a) Psoas major muscle
   b) Iliacus muscle
3. Go to depth **layer 307**. Identify the following:
   a) Piriformis muscle (Exits Pelvis via what opening?)
   b) Obturator externus muscle (Go one level deeper to remove muscle and see Obturator membrane filling in the Obturator foramen.)
4. Change to depth **layer 312**. Identify the following:
   a) Obturator membrane (Covers Obturator foramen of the pelvis.)
   b) Quadratus femoris muscle

Medial View of Hip Muscles

1. Use the **View** button to change to a Medial view of the hip region. Click **Normal Mode** button.
2. Go to depth **layer 88**. Identify the following:
   a) Iliac crest
   b) Superior pubic ramus
   c) Psoas major muscle
   d) Iliacus muscle
3. Change depth to **layer 103**. Identify the following:
   a) Sacrum
   b) Ischial spine
   c) Sacrospinous ligament and Sacrotuberous ligament
   d) Greater sciatic notch and Lesser sciatic notch (Foramen)
   e) Piriformis muscle (Exits via what Foramen?)
   f) Obturator internus muscle (Exits via what Foramen?)
4. Go to depth **layer 127** to review bony landmarks of the Pelvis. Close window and exit *ADAM Interactive* when finished.

LABORATORY 3

Muscles of the Thigh

Section I: *<u>AnatLab</u>* – Do chapter for <u>Muscles of thigh</u>
Section II: *<u>ADAM Interactive</u>* <u>– Bones of the Thigh</u>

Anterior View of the Thigh

1. Open *ADAM Interactive:*
   a) Select **Dissectible Anatomy**
   b) Select **Anterior** view and **Open**
2. Center image over the <u>knee area</u>. Use the **Depth Bar** and go to **layer 329**. Identify the following:
   a) Femur:
      1) Medial and lateral epicondyles
      2) Medial and lateral condyles
      3) Patellar surface
      4) Adductor tubercle
   b) Tibia:
      1) Medial and Lateral condyles
      2) Tibial tuberosity
      3) Fibular facet (where Head of Fibula articulates with Tibia; *not labeled*)
   c) Fibula – Head
   d) Interosseous membrane (uniting Tibia and Fibula)
3. Change depth to **layer 293**. Identify the Patella. Go to **layer 291** to see relationship of Patella to tendon of Rectus femoris muscle (part of the Quadriceps femoris muscle).

Posterior View of the Lower Limb

1. Click **View** button and change orientation to **Posterior**.
   a) Click button for **Normal Mode**.
   b) Reposition screen to show the <u>posterior thigh</u>.
2. Change depth to **layer 185**. Identify the Linea aspera on the Posterior femur.
3. Reposition the screen to the posterior <u>knee region</u> (**layer 185**). Identify the following:
   a) Femur:
      1) Adductor tubercle
      2) Medial and Lateral condyles
      3) Medial and Lateral epicondyles
   b) Tibia:
      1) Medial and Lateral condyles
      2) Fibular facet (*not labeled*; for articulation with Head of Fibula)

c) Fibula:
   1) Head

RADIOGRAPHIC VIEWS OF THE LOWER LIMB

Atlas Anatomy

1. Pull down **File** menu:
   a) **Open** – **Content** – **Atlas Anatomy** – **System** – **Skeletal**
   b) Scroll down and open **Flexed Knee (Lat)**
   c) Use **Zoom** button to reduce size of image
2. Identify the following:
   a) Femur – Condyle, Epicondyle
   b) Tibia – Condyle, Tibial tuberosity
   c) Fibula
   d) Patella (Note that the Patella articulates only with the Distal femur.)
3. Close Flexed Knee window when finished.

Section III: *ADAM Interactive* – Muscles of the Thigh

Anterior View of the Thigh

1. Open *ADAM Interactive:*
   a) Select **Dissectible Anatomy**
   b) Select **Anterior** view and **Open**
2. Center image over Anterior thigh region.
3. Use **Depth Bar** to remove skin and subcutaneous layers. Identify the Fascia lata of the Thigh.
4. Change depth to **layer 181**. Identify the following:
   a) Quadriceps femoris muscle:
      1) Rectus femoris muscle
      2) Vastus medialis muscle
      3) Vastus lateralis muscle
      4) Patellar ligament
   b) Inguinal ligament
   c) Sartorius muscle
   d) Gracilis muscle
   e) Pectineus muscle
   f) Adductor longus muscle
5. Change depth to **layer 192**. Identify the Vastus intermedius portion of Quadriceps femoris muscle.
6. Continue down to **layer 283** to remove the Vastus lateralis and medialis muscle parts.
7. At depth **layers 280** and **300**, identify the following:
   a) Pectineus muscle
   b) Adductor longus muscle
   c) Gracilis muscle
   d) Adductor brevis muscle

   e) Hamstring part of Adductor magnus muscle (labeled as Ischiocondylar part of adductor magnus)
8. At depth **layer 303**, identify Adductor brevis muscle.
9. At depth **layer 307**, identify the following:
   a) Hamstring and Adductor parts of Adductor magnus muscle
   b) Adductor hiatus (*not labeled*) – for passage of Femoral artery and vein between Thigh and Popliteal fossa regions
   c) Obturator externus muscle – passes posterior to Hip joint

Lateral View of the Thigh

1. Use **View** button to change to Lateral view of thigh region. Click **Normal Mode** button.
2. At depth **layer 60**, identify the following:
   a) Sartorius muscle
   b) Rectus femoris portion of Quadriceps femoris muscle
   c) Vastus lateralis portion of Quadriceps femoris muscle
3. At depth **layer 101**, note origins of Sartorius muscle (Anterior superior iliac spine) and Rectus femoris muscle (Anterior inferior iliac spine).
4. At depth **layer 102**, note origin of Vastus intermedius from Anterior femur.

Medial View of the Thigh

1. Use **View** button to change to Medial view of thigh region. Click **Normal Mode** button.
2. At depth **layer 8**, identify Sartorius muscle, Gracilis muscle, and Semitendinosus muscle. (Note how all three of these muscles insert together onto the anterior-medial side of the Proximal tibia. What is the name for this three-part insertion?)
3. At depth **layer 20**, identify the following:
   a) Hamstring part of Adductor magnus muscle (labeled as Ischiocondylar part)
   b) Adductor hiatus (*not labeled*) – for passage of Femoral artery and vein between Thigh and Popliteal fossa regions

Posterior View of the Thigh

1. Use **View** button to change to Posterior view of thigh region. Click **Normal Mode** button.
2. Go to both depth **layers 11** and **113**. Identify the following:
   a) Gluteus maximus (layer 11 only)
   b) Biceps femoris muscle – short and long heads
   c) Semimembranosus muscle
   d) Semitendinosus muscle
   e) Hamstring part of Adductor magnus muscle (labeled as Ischiocondylar part)
   f) Adductor part of Adductor magnus muscle (layer 113 only)
   g) Gracilis muscle
   h) Vastus lateralis
3. At depth **layer 116**, identify the Semimembranosus muscle.
4. At depth **layer 121**, identify the following:
   a) Adductor magnus muscle (Adductor and Hamstring parts)
   b) Short head of Biceps femoris muscle
   c) Adductor hiatus – for passage of Popliteal artery and vein
5. At depth **layer 122**, identify the Adductor longus muscle and Adductor brevis muscle.

6. At depth **layer 136**, note the origins of Vastus medialis and Vastus lateralis from Linea aspera of Posterior femur.

Thigh Muscles – Practice and Review

1. Practice your understanding of muscles of the thigh using pinned views of the lower limb.
   a) Pull down **File** menu.
   b) **Open – Content – Atlas Anatomy – System – Muscular**
   c) Select from the following:
      1) **Anterior Thigh** – dissected cadaver limb
      2) **Dissection of Posterior Thigh** – dissected cadaver limb
      3) **Medial Thigh** – drawing
      4) **Posterior Thigh** – drawing

LABORATORY 4

Muscles of the Leg & Foot

Section I: ***AnatLab*** – Do chapter for Muscles of leg & foot
Section II: ***ADAM Interactive*** **– Bones of the Leg and Foot**

1. Pull down **File** menu – **Open** – **Atlas Anatomy** – **Region** – **Lower Limb**. Select **Bones of the Leg and Foot (Ant)**. Identify the following:
   a) Body of fibula
   b) Body of tibia
   c) Calcaneus bone
   d) First and fifth metatarsal bones
   e) Interosseous membrane of leg
   f) Lateral malleolus of fibula
   g) Medial malleolus of tibia
   h) Patella
   i) Proximal, middle, & distal phalanx bones of toes
   j) Talus bone
2. **File** menu – **Open** – **Atlas Anatomy** – **Region** – **Lower Limb**. Select **Bones of the Leg and Foot (Post)**. Identify the following:
   a) Body of fibula
   b) Body of tibia
   c) Calcaneus bone
   d) Femur
   e) First and fifth metatarsal bones
   f) Proximal, middle, & distal phalanx bones of toes
   g) Talus bone

Anterior View of the Lower Limb

1. Open Dissectible Anatomy window:
   a) Pull down **File** menu – select **Open** – select **Content**
   b) Click tab for **Dissectible Anatomy**
   c) Select **Anterior** view and **Open**
2. Move to the knee area and go to **layer 329**. Identify the following:
   a) Femur:
      1) Medial and lateral epicondyles
      2) Medial and lateral condyles
   b) Tibia:
      1) Medial and Lateral condyles
      2) Fibular facet (where Head of fibula articulates with Tibia; *not labeled*)

   c) Fibula – Head
   d) Interosseous membrane (uniting Tibia and Fibula)
3. Center the image of the ankle area and change depth to **layer 329**. Identify the following:
   a) Tibia:
      1) Medial malleolus
      2) Fibular notch (for articulation with Fibula; *not labeled*)
   b) Fibula:
      1) Lateral malleolus
      2) Facet for talus (for articulation with Tibia; *not labeled*)
   c) Tarsal bones of foot:
      1) Talus
      2) Calcaneus (only partially visible)
      3) Navicular
      4) Cuboid
      5) Medial, Intermediate, and Lateral Cuneiform bones
   d) Metatarsal bones – first (for great toe) to fifth (for little toe)
   e) Proximal and distal phalanges (phalanx bones) of great toe (digit 1)
   f) Proximal, middle, and distal phalanges of digits 2-5
4. Change depth to **layer 330**. Note box shape of ankle articulation (between tibia, fibula, and talus).

Posterior View of the Lower Limb

1. Click **View** button and change orientation to **Posterior**.
   a) Click button for **Normal Mode**.
   b) Reposition screen to show Posterior hip and pelvic regions.
2. Position the screen to the knee region and go to **layer 185**. Identify the following:
   a) Femur:
      1) Medial and Lateral condyles
      2) Medial and Lateral epicondyles
   b) Tibia:
      1) Medial and Lateral condyles
      2) Fibular facet (*not labeled*; for articulation with Head of Fibula)
      3) Soleal line
   c) Fibula:
      1) Head
3. Reposition to ankle and foot area. (**layer 185**). (*View shows plantar surface of the foot.*) Identify the following:
   a) Tibia – Medial malleolus
   b) Fibula – Lateral malleolus
   c) Calcaneus bone: Sustentaculum tali (*not labeled*) – Medial extension of Calcaneus bone (with Groove for tendon of Flexor digitorum longus muscle
4. Click **View** button and change to a Lateral view of the ankle and foot. Click **Normal Mode** button.
5. Move depth bar to **layer 287**. Identify the following:
   a) Tibia
   b) Fibula – Lateral malleolus
   c) Calcaneus bone
   d) Talus bone
   e) Cuboid bone

   f) Navicular bone
   g) Lateral and intermediate cuneiform bones
   h) Metatarsal and phalanx bones
6. When finished, close the window containing the Lateral view.

Radiographic Views of the Lower Limb

Atlas Anatomy

1. Pull down **File** menu:
   a) **Open – Content – Atlas Anatomy – System – Skeletal**
   b) Scroll down and open **Foot (Lat)**
   c) Use **Zoom** button to reduce size of image
2. Identify:
   a) Tibia and Fibula
   b) Calcaneus, Talus, Navicular bones
   c) Metatarsal and Phalanx bones
3. Close window when finished.

Section III: *ADAM Interactive* – Muscles of the Leg and Foot

Anterior View of the Leg and Foot

1. Open *ADAM Interactive:*
   a) Select **Dissectible Anatomy**
   b) Select **Anterior** view and **Open**
2. Center image over Anterior leg region.
3. Use **Depth Bar** to remove skin and subcutaneous layers. Identify the Crural Fascia of the Leg.
4. Change depth to **layer 182**. Identify the following:
   a) Anterior border of Tibia
   b) Extensor retinaculum
5. Change depth to **layer 186**. Identify the following:
   a) Tibialis anterior muscle and tendon
   b) Extensor digitorum longus muscle and tendon
   c) Tendon of Fibularis tertius muscle
   d) Tendon of Extensor hallucis longus muscle
6. Change depth to **layer 187** and identify Extensor hallucis longus muscle.
7. Change depth to **layer 284**. Identify the following:
   a) Fibularis tertius muscle
   b) Extensor digitorum brevis muscle and Extensor hallucis brevis muscle on dorsal foot

Lateral View of the Leg

1. Use **View** button to change to Lateral view of leg region. Click **Normal Mode** button.
2. Go to depth **layer 82**. Identify the Fibular retinaculum.
3. Go to depth **layer 87**. Identify the following:
   a) Fibularis longus muscle
   b) Tendon of Fibularis brevis muscle

   c) Tibialis anterior muscle
   d) Extensor digitorum longus muscle
   e) Fibularis tertius muscle
4. Go to depth **layer 88** and identify the Fibularis brevis muscle.

Posterior View of the Leg

1. Use **View** button to change to Posterior view of leg region. Click **Normal Mode** button.
2. Go to depth **layer 8**. Identify the following:
   a) Gastrocnemius muscle
   b) Calcaneal tendon
3. Go to depth **layer 137**. Identify the following:
   a) Plantaris muscle
   b) Soleus muscle
4. Go to depth **layer 153**. Identify the following:
   a) Popliteus muscle (seen better at **layer 161**)
   b) Flexor digitorum longus muscle
   c) Tibialis posterior muscle
   d) Flexor hallucis longus muscle

View of the Plantar Foot

1. While still in Posterior view orientation, return to depth **layer 0** and center the image over the Plantar foot region.
2. Go to depth **layer 8** and identify the Plantar aponeurosis.
3. Go to depth **layer 136**. Identify the following:
   a) Abductor hallucis muscle
   b) Flexor digitorum brevis muscle
   c) Abductor digiti minimi muscle
4. Go to depth **layer 149**. Identify the following:
   a) Tendon of Flexor hallucis longus muscle
   b) Tendon of Flexor digitorum longus muscle
   c) Quadratus plantae muscle
5. Go to depth **layer 155** and identify the insertion of the Tendon of Tibialis posterior muscle.
6. Go to depth **layer 173** and identify the insertion of the Tendon of Fibularis (Peroneus) longus muscle.

Medial View of the Leg

1. Use **View** button to change to Medial view of ankle region. Click **Normal Mode** button.
2. Go to depth **layer 8**. Identify the following:
   a) Flexor retinaculum
   b) Gastrocnemius muscle
   c) Soleus muscle
   d) Calcaneal tendon
   e) Abductor hallucis muscle
   f) Plantar aponeurosis

3. Go to depth **layer 62**. Identify the following:
   a) Flexor hallucis longus muscle and tendon
   b) Flexor digitorum longus muscle and tendon
   c) Tendon of Tibialis posterior muscle
   d) Tendon of Tibialis anterior muscle
   e) Sustentaculum tali of calcaneus bone (small area of calcaneus bone between tendons of tibialis posterior and flexor digitorum longus)

LABORATORY 5

Introduction to Nervous System

Section I: ***ADAM Interactive*** **– Introduction to the Nervous System**

1. Open *ADAM Interactive:*
   a) Select **Atlas Anatomy**
   b) Select **System**
   c) Scroll down and select **Nervous** system
   d) Scroll down and select **Nerves of Thoracic Wall**
2. Use the pull-down menu at the top of the image, or click on the pin heads to locate and identify the following:
   a) Spinal cord
   b) Dorsal and Ventral roots of Spinal nerve
   c) Dorsal root ganglion
   d) Spinal nerve (Intercostal nerve)
   e) Cutaneous (sensory) branches of Intercostal nerve
3. Close the Nerves window when finished.
4. Pull down **File** menu – **Open** – **Atlas Anatomy** – **System** – **Nervous**. Select **Dermatomes of Trunk (Ant)**. Identify the following:
   a) Anterior and Lateral cutaneous nerve branches
   b) Dermatomes of Cervical, Thoracic, Lumbar, and Sacral spinal nerves
5. Close window when finished.
6. Pull down **File** menu – **Open** – **Atlas Anatomy** – **System** – **Nervous**. Select **Dermatomes of Trunk (Post)**. Identify the following:
   a) Cutaneous nerve branches of spinal nerves
   b) Lateral cutaneous nerve branches of spinal nerves
   c) Dermatomes of Cervical, Thoracic, Lumbar, and Sacral spinal nerves
7. Close window when finished.
8. Pull down **File** menu – **Open** – **Atlas Anatomy** – **Region** – **Lower Limb**. Select **Dermatomes of Lower Limb (Ant)**. Identify Dermatomes of Lumbar and Sacral spinal nerves.
9. Close window when finished.
10. Pull down **File** menu – **Open** – **Atlas Anatomy** – **Region** – **Lower Limb**. Select **Dermatomes of Lower Limb (Post)**. Identify Dermatomes of Lumbar and Sacral spinal nerves.
11. Close window when finished.

Posterior View of the Spinal Cord and Spinal Nerves

1. Open Dissectible Anatomy window.
   a) Pull down **File** menu – select **Open** – select **Content**
   b) Click tab for **Dissectible Anatomy**
   c) Select **Posterior** view and **Open**
2. Center image over the back.
3. Use the **Depth Bar** and go to depth **layer 180** (if image disappears or becomes black/white, press **Normal Mode** button). Identify the following:
   a) Spinal cord
   b) Dorsal nerve roots and Ventral nerve roots
   c) Dorsal root ganglia
   d) Note that the spinal nerves exit between adjacent vertebrae and are named (numbered) in relation to the vertebrae.

LABORATORY 6

Nerves and Blood Vessels of the Lower Limb

Section I: *AnatLab* – Do chapters for Nerves of lower limb and Blood vessels of lower limb
Section II: *ADAM Interactive* – Blood Vessels and Nerves of the Hip and Thigh Regions

Anterior View of the Hip and Thigh Region

1. Open Dissectible Anatomy window.
   a) Pull down **File** menu – select **Open** – select **Content**
   b) Click tab for **Dissectible Anatomy**
   c) Select **Posterior** view and **Open**
2. Position the image for an Anterior view of the hip and superior thigh region.
3. Go to depth **layer 5**. Identify the following:
   a) Great (greater) saphenous vein
4. Go to depth **layers 18 and 181**. Identify the following:
   a) Borders of femoral triangle: Inguinal ligament (superior border), Sartorius muscle (lateral border), Adductor longus muscle (medial border)
   b) Contents of Femoral triangle: Femoral nerve, Femoral artery, and Femoral vein
   c) Anterior branch of Obturator nerve (Located anterior to what structure?)
5. Change depth to **layer 189**. Identify the Femoral artery running within Adductor canal (*not labeled*) – deep to Sartorius muscle, in groove between Vastus medialis muscle and Adductor muscles.
6. Go to depth **layer 240**. Identify the following:
   a) External iliac artery and vein – superior to line of inguinal ligament
   b) Femoral artery and vein – inferior to line of inguinal ligament
   c) Deep femoral artery and 1st Perforating artery
   d) Lateral circumflex artery
7. Go to depth **layer 268** and identify these Lumbar and Sacral plexus components within the Pelvis:
   a) L1 to S4 Spinal nerves
   b) Femoral nerve (back up to **layer 266** to see relationship of nerve to Iliacus and Psoas major muscles within Pelvic region)
   c) Obturator nerve
   d) Lumbosacral trunk (formed by L4 plus L5 spinal nerves; contributes to Sacral plexus)
   e) Sciatic nerve
8. Change to depth **layer 282**. Identify the following:
   a) Femoral artery and vein in Adductor canal region
   b) Adductor hiatus – formed by gap between Adductor and Hamstring (Ischiocondylar) parts of Adductor magnus muscle

9. Go to depth **layers 302 and 305**. Identify the following:
   a) Adductor brevis muscle
   b) Anterior and posterior branches of Obturator nerve
   c) Medial circumflex femoral artery
10. Change to depth **layer 311**. Identify the Obturator artery and vein (Exits Pelvis via what opening?)

Medial View of Pelvis and Lower Limb

1. Use the **View** button to change to a Medial view of the lower limb. Click **Normal Mode** button.
2. Change depth to **layer 4**. Identify the Greater saphenous vein running superficial to the Deep investing fascias of the Thigh (Fascia lata) and Leg (Crural fascia). Follow the entire length of the Great saphenous vein as it passes up the medial leg and thigh. Also note its origin from the Dorsal venous arch of the Foot.
3. Go to depth **layer 22**. Identify the following in the Thigh:
   a) Femoral artery and vein passing into Adductor hiatus (*not labeled*)
   b) Deep femoral artery, with Perforating arteries wrapping around to posterior side of Femur
4. Go to depth **layer 65**. Identify in the Pelvic region the External iliac artery and Internal iliac artery.
5. In depth **layers 86 and 89**, identify the following in the Pelvis:
   a) Femoral nerve; note relationship to Iliacus and Psoas major muscles
   b) Obturator nerve; exits Pelvis with Obturator artery and vein at superior border of Obturator internus muscle
   c) L4-S4 Spinal nerves and the Lumbosacral trunk
   d) Sciatic nerve

Posterior View of Hip and Thigh Regions

1. Use **View** button to change to a Posterior view of the hip region. Click **Normal Mode** button.
2. Go to depth **layers 104-105**. Identify the following:
   a) Sciatic nerve
   b) Superior gluteal nerve, artery, and vein
   c) Inferior gluteal nerve, artery, and vein
3. Adjust image to now show the distal thigh and posterior knee regions. Identify the borders of the popliteal fossa region:
   a) Biceps femoris muscle – superior-lateral border of popliteal fossa
   b) Semitendinosus muscle – superior-medial border
   c) Medial and lateral head of Gastrocnemius muscle – medial-inferior and lateral-inferior borders
4. Go to **depth layers 120 and 121**. Identify the following:
   a) Sciatic nerve
   b) Tibial nerve
   c) Common fibular nerve
   d) Deep femoral artery and Perforating arteries (also seen at **layer 132**)
   e) Popliteal artery emerging through Adductor hiatus (*not labeled*)

Section III: *ADAM Interactive* – Blood Vessels and Nerves of the Leg and Foot

Posterior View of Leg and Plantar Foot

1. Adjust position of image to show the Posterior leg region.
2. Change depth to **layer 3**. Identify the Small Saphenous vein. Note its relationship to the Crural fascia as it passes up the posterior leg. The Small saphenous vein drains into what vein?
3. Go to **layers 137 to 141** and note positions of the Tibial nerve and Popliteal artery and vein to the Plantaris and Soleus muscles.
4. Go to **layers 140 to 142**. Identify the following:
   a) Tibial nerve and Posterior tibial artery – follow these into the Plantar foot. Note relationships to Medial malleolus and Abductor hallucis muscle.
   b) Medial plantar artery & nerve and Lateral plantar artery & nerve
   c) Fibular artery and Anterior tibial artery (better seen at **layer 156**)

Lateral View of Leg and Foot

1. Use **View** button to change to a Lateral view of the leg. Click **Normal Mode** button.
2. Go to depth **layer 83**. Identify the following:
   a) Common fibular nerve; note relationship to Head of Fibula
   b) Superficial fibular nerve
3. Go to depth **layer 265**. Identify the following:
   a) Deep fibular nerve
   b) Tibial artery
   c) Anterior tibial artery; follow across ankle to dorsum of foot, where name changes to Dorsalis pedis artery
   d) Posterior tibial artery
   e) Fibular artery

Anterior View of Leg and Foot

1. Use **View** button to change to an Anterior view of the leg and foot. Click **Normal Mode** button.
2. Go to **layer 296**. Identify the following:
   a) Deep fibular nerve
   b) Anterior tibial artery; follow this onto the Dorsal foot where it becomes the Dorsalis Pedis artery

Medial View of Leg and Foot

1. Use **View** button to change to a Medial view of the leg and foot. Click **Normal Mode** button.
2. Go to **layer 8** and identify the Flexor retinaculum.
3. Go to **layer 32**. At the Medial malleolus, identify the Tibial nerve and Posterior tibial artery. Note the relationship of these to the muscle tendons passing from the Deep posterior compartment of the Leg into the Plantar foot.
4. In the plantar foot (**level 32**), identify the Medial plantar and Lateral plantar nerve and artery.

LABORATORY 7

Joints of the Lower Limb

Section I: ***AnatLab*** – Do chapter for Review/Quiz of Lower Limb
Section II: ***ADAM Interactive* – Knee Joint**

1. Open *ADAM Interactive:*
   a) Select **Dissectible Anatomy**
   b) Select **Anterior** view and **Open**
2. Center image over Anterior knee region.
3. Change depth to **layer 188** and identify Patella (*not labeled*) and Patellar ligament (Associated with what muscle?).
4. Go to depth **layer 293**. Identify the following:
   a) Fibular collateral ligament
   b) Tibial collateral ligament
5. Go to depth **layer 321**. Identify the following:
   a) Medial and Lateral femoral condyles with articular cartilage
   b) Medial and Lateral tibial condyles with articular cartilage
   c) Medial and Lateral meniscus
   d) Anterior cruciate ligament
6. Change to a Posterior view of the knee. Click **Normal Mode** button.
7. Go to depth **layers 160 and 161**. Identify the following:
   a) Fibrous capsule of Knee joint
   b) Synovial capsule of Knee joint
   c) Tibial and Fibular collateral ligaments
   d) Popliteus muscle
   e) Posterior cruciate ligament
8. At depth **layer 170**, identify the following:
   a) Anterior and Posterior cruciate ligaments
   b) Medial and Lateral meniscus
   c) Articular cartilages of Femoral and Tibial condyles
9. Change to a Lateral view of the knee. Click **Normal Mode** button.
10. Go to depth **layer 82** and identify the Iliotibial tract. Note how this inserts onto the Tibia below the Lateral knee. This attachment site is anterior to the Axis of knee bending. What action does pulling of the Iliotibial tract have on the Knee joint?
11. Go to depth **layer 266**. Identify the following:
    a) Fibular collateral ligament
    b) Popliteus muscle
    c) Fibrous capsule of Knee joint
12. Go to depth **layer 283**. Identify the following:
    a) Lateral meniscus
    b) Articular cartilages of Femur and Tibia condyles

   c) Patella
13. Go to depth **layer 288** and identify the Intercondylar eminence of the Tibia.
14. Change to a Medial view of the knee. Click **Normal Mode** button.
15. Go to depth **layer 33**. Identify the Tibial collateral ligament and the Fibrous capsule of the Knee joint.
16. Go to depth **layer 120**. Identify the following:
    a) Medial meniscus
    b) Articular cartilages of Femur and Tibia condyles
    c) Patella

Atlas Anatomy of the Knee

1. **File – Open – Content – Atlas Anatomy – System – Skeletal system**
2. Open **Articular Surface of the Tibia** and identify the following:
    a) Medial and lateral condyles of the Tibia
    b) Medial and lateral meniscus
    c) Anterior and posterior cruciate ligaments (Note that these are named for their attachment sites on the Tibia.)
    d) Fibular and tibial collateral ligaments
    e) Ligament of popliteus muscle

LABORATORY 8

Bones of the Skull

Use the <u>AnatLab</u> and <u>ADAM Interactive</u> software programs as you identify the bones and bony landmarks of the skull that are given in your course pack notes. Also <u>use the skulls</u> available in the computer lab. Landmarks of the skull <u>cannot</u> be learned solely by looking at pictures.

Section I: *<u>AnatLab</u>* – Do chapter for <u>Bones of skull</u>
Section II: *<u>ADAM Interactive</u>* <u>– Bones of the Skull</u>

Lateral View of the Skull

1. **File – Open – Content – Dissectible Anatomy**
2. Open a <u>Lateral view of the head</u>. Go to depth **layer 118** and identify the following:
   a) Frontal, Parietal, Occipital, Nasal, and Maxilla bones
   b) Temporal bone – Mastoid process, External auditory meatus, Zygomatic process
   c) Zygomatic bone – Temporal process
   d) Zygomatic arch = Zygomatic process of temporal bone + Temporal process of zygomatic bone
   e) Greater wing of Sphenoid bone
   f) Mandible – Condyle (Head), Neck, Ramus, Coronoid process, Body, Angle
   g) Coronal, Lambdoid, Squamosal sutures
3. At depth **layer 187**, identify the following:
   a) Styloid process of the Temporal bone
   b) Mandibular fossa and Articular tubercle of Temporal bone
4. Go to depth **layer 208** and identify the following:
   a) Frontal and Maxillary sinuses
   b) Sphenoid, Lacrimal, Ethmoid, Nasal, and Maxilla bones
   c) Lateral pterygoid plate
   d) Sella turcica (*not labeled*; bony parts of Sphenoid bone that surround Hypophyseal fossa)
   e) Occipital condyle and margin of Foramen magnum
5. At depth **layer 209**, identify the following:
   a) Frontal and Sphenoid sinuses
   b) Ethmoid bone – Crista galli, Perpendicular plate
   c) Nasal septum: Perpendicular plate of Ethmoid bone, Vomer bone, and Nasal septal cartilage
   d) Parts of the hard palate: Palatine process of Maxilla and Palatine bone
   e) Sella turcica
6. At depth **layer 237**, identify the following:
   a) Superior, Middle, and Inferior nasal concha
   b) Hyoid bone

Anterior View of Skull

1. Open an Anterior view of the head region. Go to depth **layer 48** and identify the following:
   a) Frontal, Nasal, Maxilla, Zygomatic, and Lacrimal bones
   b) Perpendicular plate of ethmoid bone (Superior part of Nasal septum)
   c) Vomer bone (Inferior part of Nasal septum)
   d) Middle and Inferior nasal concha
   e) Mandible – Body, Mental foramen
2. Go to depth **layer 49**. Identify the following:
   a) Maxillary sinus and Ethmoid air cells
   b) Middle and inferior nasal conchae

Posterior View of the Skull

1. Open a Posterior view of the head region.
2. Go to depth **layer 100** and identify the following:
   a) Occipital bone – External occipital protuberance, Superior nuchal line
   b) Parietal bone
   c) Sagittal and Lambdoid sutures

Atlas Anatomy of the Skull

1. **File – Open – Content – Atlas Anatomy – System – Skeletal system**
2. Open **Skull (Inf)** and identify the following:
   a) Lateral and Medial pterygoid plates
   b) Vomer bone
   c) Mandibular fossa of Temporal bone
   d) Occipital condyle
   e) Styloid process
   f) Palatine process of Maxilla bone
   g) Palatine bone (labeled as Horizontal plate of Palatine bone)

1. **File – Open – Content – Atlas Anatomy – System – Skeletal system**
2. Open **Skull (Lat) 1** and identify the following:
   a) Frontal, Parietal, Occipital, Nasal, Lacrimal, and Maxilla bones
   b) Temporal bone – Mastoid process, Zygomatic process, Mandibular fossa, Articular tubercle
   c) Zygomatic bone – Temporal process
   d) Zygomatic arch = Zygomatic process of temporal bone + Temporal process of zygomatic bone
   e) Greater wing of Sphenoid bone
   f) Mandible – Condyle (Head), Neck, Ramus, Coronoid process, Mandibular notch, Body, Angle
   g) Coronal, Lambdoid, Squamosal sutures

1. **File – Open – Content – Atlas Anatomy – System – Skeletal system**
2. Open **Muscle Atts – Skull (Med)** and identify the following:
   a) Frontal and Sphenoid sinuses
   b) Mylohyoid line (red line on mandible for attachment of Mylohyoid muscle)

1. **File – Open – Content – Atlas Anatomy – System – Skeletal system**
2. Open **Muscle Atts – Skull (Inf)** and identify the following:
   a) Mandible – Body, Lingula
   b) Lateral and Medial pterygoid plates
   c) Vomer bone
   d) Mandibular fossa of Temporal bone
   e) Occipital condyle
   f) Zygomatic arch = Zygomatic process of temporal bone + Temporal process of zygomatic bone

1. **File – Open – Content – Atlas Anatomy – System – Skeletal system**
2. Open **Cranial Cavities (Sup) 2** and identify the following:
   a) Anterior, Middle, and Posterior cranial cavities
   b) Lesser wing of Sphenoid bone
   c) Petrous part of Temporal bone (posterior margin forms Petrous ridge – *not labeled*)
   d) Hypophyseal fossa (for pituitary gland) surrounded by bone of Sella turcica (*not labeled*)

1. **File – Open – Content – Atlas Anatomy – System – Skeletal system**
2. Open **Mandible (Post)** and identify the following:
   a) Condyle, Neck, Ramus, Angle, Body
   b) Mylohyoid line (*not labeled*; red line for attachment of Mylohyoid muscle)
   c) Mandibular foramen and Lingula (*not labeled*; small flap of bone next to Mandibular foramen)

Other *Atlas Anatomy* views that may be helpful for the identification of skull bones:
Walls of Orbit (Ant)
Sagittal Section of Head and Neck (cadaver image)

ADAM Interactive – 3D Anatomy of the Skull

1. **File – Open – Content – 3D Anatomy – 3D Skull**
   a) Use <u>arrow buttons</u> to rotate skull
   b) Use **Zoom** button to increase-decrease view of the skull
2. Use <u>pull-down menu</u> to select and identify individual bones and bony landmarks.

LABORATORY 9

The Brain

Section I: *AnatLab* – Do chapter for The Brain
Section II: *ADAM Interactive* – The Brain

Lateral View of the Brain

1. **File – Open – Content –Dissectible Anatomy**
2. Open a Lateral view of the head. Go to depth **layer 191** and identify the following:
   a) Gyri of the Frontal, Parietal, Occipital, and Temporal lobes
   b) Central sulcus (Separates which 2 lobes?)
   c) Precentral gyrus, Postcentral gyrus (What specific functional brain areas occupy these gyri?)
   d) Lateral sulcus (Lateral fissure)
   e) Parieto-occipital sulcus
   f) Cerebellum
3. Go to depth **layer 288**. Identify the following:
   a) Frontal, Parietal, and Occipital lobes
   b) Parieto-occipital sulcus
   c) Calcarine sulcus (anterior-posterior running sulcus in middle of occipital lobe)
   d) Cingulate gyrus
   e) Corpus callosum
   f) Thalamus (forming lateral wall of third ventricle) and Massa intermedia (labeled as Interthalamic adhesion)
   g) Hypothalamus (forming lateral wall of third ventricle) and Mammillary bodies
   h) Pineal body (Pineal gland)
   i) Pituitary gland (divided into Adenohypophysis and Neurohypophysis)
   j) Optic chiasm
   k) Midbrain with Superior colliculus, and Inferior colliculus
   l) Pons
   m) Medulla oblongata, with Olives (*not labeled*; oval-shaped surface bumps on lateral side of medulla)
   n) Cerebellum (note tightly folded surface layer of gray matter and highly branched white matter forming the Arbor vitae, *not labeled*)

Anterior View of the Brain

1. **File – Open – Content –Dissectible Anatomy**
2. Open Anterior view of the head. Go to depth **layer 49** and identify the Frontal lobe. Note that gray matter (*not labeled*; darker layer) forms the surface of the brain, with white matter located deeper. Also note how the gyri (*folds*) and sulci (*grooves*) serve to increase

the brain's surface area and thus the amount of gray matter. (Define gray matter and white matter; i.e., at the cellular level, what structures form gray matter and white matter?)

3. Go to depth **layer 64** and identify the Frontal lobe and Temporal lobe.
4. Go to depth **layer 330** and identify the following:
   a) Cingulate gyrus
   b) Basal nuclei (labeled as its individual nuclei: the Caudate, Putamen, and Globus pallidus)
   c) Thalamus
   d) Corpus callosum
   e) Pons

Atlas Anatomy of the Brain

1. **File – Open – Content – Atlas Anatomy – Region – Head & Neck**
2. Open **Brain (Lat)** and identify the following:
   a) Frontal, Parietal, Occipital, and Temporal lobes
   b) Central sulcus, Parieto-occipital sulcus, Lateral sulcus (Lateral fissure)
   c) Precentral gyrus, Postcentral gyrus
   d) Pons, Medulla oblongata, Olive
   e) Cerebellum
   f) Spinal cord
3. Close window when finished.

1. **File – Open – Content – Atlas Anatomy – Region – Head & Neck**
2. Open **Sagittal Section of Head & Neck** (cadaver image) and identify the following:
   a) Corpus callosum
   b) Optic chiasm
   c) Pituitary gland
   d) Thalamus, Midbrain, Pons, Medulla oblongata, and Cerebellum
   e) Spinal cord and Foramen magnum
3. Close window when finished.

1. **File – Open – Content – Atlas Anatomy – Region – Head & Neck**
2. Open **Base of Brain (Inf)** and identify the following:
   a) Frontal, Temporal, and Occipital lobes
   b) Pituitary gland
   c) Mammillary bodies (part of Hypothalamus)
   d) Pons
   e) Cerebellum
   f) Medulla oblongata (*not labeled*) with Pyramid, Olive
   g) Spinal cord

3D Anatomy of the Brain

1. **File – Open – Content – 3D Anatomy**
2. Open **3D** Brain. Use the pull-down menu to identify the following:
   a) Parts of Cerebrum:
      1) Frontal, Parietal, Occipital, and Temporal lobes
      2) Central sulcus (Separates which lobes?)
      3) Precentral gyrus and Postcentral gyrus (What functional brain regions are associated with each of these?)

    4) Parieto-occipital sulcus and Lateral sulcus (Lateral fissure)
    5) Corpus callosum (Function?)
    6) Basal nuclei (areas of deep gray matter) – Caudate, Putamen, and Globus pallidus
    7) Limbic structures (areas of deep gray matter) – Amygdaloid body (amygdala), Hippocampus

b) Parts of Diencephalon:
    1) Thalamus – connected across midline by Massa intermedia (Interthalamic adhesion)
    2) Mammillary bodies (part of the Hypothalamus) and Pituitary gland (connected to Hypothalamus)
    3) Pineal body (gland)

c) Parts of Midbrain:
    1) Midbrain
    2) Superior colliculus and Inferior colliculus (Functions of each?)

d) Cerebellum

e) Parts of Brainstem:
    1) Pons
    2) Medulla oblongata – Pyramid and Olive

LABORATORY 10

Ventricles, Meninges, and Circulation of the Brain

Section I: ***<u>AnatLab</u>*** – Do chapters for <u>Ventricles and Cerebrospinal Fluid</u> and <u>Meninges/Circulation</u>
Section II: ***<u>ADAM Interactive</u>***

3D Anatomy of the Brain

1. **File – Open – Content – 3D Anatomy – 3D Brain**
2. Use the <u>pull-down menu</u> at the top of the page to identify the following structures:
   a) Frontal, Occipital, Parietal, Temporal lobes
   b) Cerebellum
   c) Dura mater, Arachnoid mater, Subarachnoid space, Pia mater
   d) Falx cerebri (listed as Cerebral falx) and Tentorium cerebelli
   e) Superior sagittal sinus, Inferior sagittal sinus, Straight sinus, Confluens of sinuses, Transverse sinus, Sigmoid sinus, Cavernous sinus
   f) Middle meningeal artery (Enters cranial cavity through what opening?)
   g) Anterior Cerebral, Anterior communicating, Basilar, Internal Carotid, Middle Cerebral, Posterior communicating, and Vertebral arteries
   h) Lateral ventricles, Third ventricle, Fourth ventricle, Interventricular foramen, Cerebral aqueduct
3. <u>Close window</u> when finished.

Dissectible Anatomy

1. Open **Dissectible Anatomy** – <u>Lateral view of head</u>. Change depth to **layer 187**. Then click down to **layer 188**. Identify the following:
   a) Dural mater and Middle meningeal artery
   b) Internal jugular vein, Sigmoid sinus, Transverse sinus, Confluence of sinuses
2. Go to depth **layer 190** and identify branches of the Middle cerebral artery.
3. Go to depth **layers 192** and **202**. Identify the following:
   a) Falx cerebri
   b) Superior sagittal, Inferior sagittal, Straight, and Cavernous sinuses
   c) Great cerebral vein, Confluence of sinuses
   d) Internal carotid, Basilar, and Anterior cerebral arteries
   e) Septum pellucidum and Choroid plexus
   f) Third ventricle, Cerebral aqueduct, and Fourth ventricle
4. Go to depth **layer 228** and identify the Basilar and Vertebral arteries.
5. Go to depth **layer 265** and identify the following:
   a) Anterior cerebral artery

   b) Posterior cerebral artery
   c) Posterior communicating artery
6. Change to an Anterior view of the head. At depth **layer 64**, identify the following:
   a) Falx cerebri
   b) Superior sagittal sinus and Inferior sagittal sinus
7. Go to depth **layer 275** and identify the following:
   a) Falx cerebri with Superior and Inferior sagittal sinuses
   b) Lateral ventricles
   c) Third ventricle

Atlas Anatomy of Blood Supply to the Brain

1. Review parts of Cerebral Arterial Circle: Open **Atlas Anatomy** – **Region** – **Head & Neck** – select **Cerebral Arterial Circle (Inf)**
2. Identify contributions to and branches of Cerebral arterial circle.
3. Close window when finished.

LABORATORY 11

Spinal Cord, Meninges, and Spinal Tracts

Section I: ***AnatLab*** – Do chapter for Spinal Cord
Section II: ***ADAM Interactive* – Spinal Cord and Meninges**

Posterior View of the Spinal Cord

1. Open **Dissectible Anatomy** – Posterior view of back. Go to depth **layer 176** and identify Spinal nerves.
2. Go to depth **layers 178-180** and identify the following:
   a) Dura mater and Arachnoid mater
   b) Dorsal root ganglia and Dorsal nerve roots
   c) Filum terminale and Cauda equina
   d) Conus medullaris
   e) Note Cervical and Lumbosacral enlargement areas of Spinal cord (*not labeled*)
3. Use **View** button to change to a Medial view of the trunk region. Click **Normal Mode** button.
4. At depth **layer 0**, identify the following:
   a) Spinal cord (note continuity with Brain stem)
   b) Conus medullaris
   c) Filum terminale and Cauda equina
5. Change depth to **layer 127** and identify the following:
   a) Dura mater
   b) Dorsal and Ventral nerve roots

Atlas Anatomy

1. **File – Open – Content – Atlas Anatomy – System – Skeletal – Sagittal Section of Lower Spine**
2. Identify the following:
   a) Spinal cord, with Conus medullaris
   b) Bodies of L1 and L2 vertebrae. Note that the Spinal cord terminates at the L1-L2 vertebral level.
   c) Filum terminale (thin cord-like structure extending from inferior tip of Spinal cord. Formed by which meningeal layer?)
   d) Dura mater, Arachnoid mater, and Subarachnoid space
   e) Cauda equina formed by descending nerve roots

LABORATORY 12

Autonomic Nervous System

Section I: ***AnatLab*** – Do chapter for Autonomic Nervous System

Section II: ***ADAM Interactive*** **– Autonomic Nervous System**

Anterior View of Autonomic Nervous System

1. Open **Dissectible Anatomy** – Anterior view of thoracic-abdominal area.
2. Change depth to **layer 264**. Scroll up and down (neck to pelvis) and identify the following:
   a) Sympathetic trunk (How far superiorly & inferiorly does this extend?)
   b) Chain (sympathetic) ganglia. These contain which specific type of neuron cell bodies?
   c) Superior cervical ganglion (in neck area)
   d) White rami communicantes and Gray rami communicantes. Specific type of nerve fibers carried by each?
   e) Greater splanchnic nerve (labeled as Thoracic splanchnic nerve)
3. Change depth to **layer 239**. Identify the following:
   a) Celiac ganglion, Superior mesenteric ganglion, Inferior mesenteric ganglion. These contain which specific type of neuron cell bodies?
   b) Greater splanchnic nerve (labeled as Thoracic splanchnic nerve). This carries which specific type of nerve fibers?

Lateral View of the Autonomic Nervous System

1. Use **View** button to change to a Lateral view of the thoracic and abdominal regions. Click **Normal Mode** button.
2. Go to depth **layer 225** and identify the following:
   a) Sympathetic trunk
   b) Chain (sympathetic) ganglia
   c) Superior cervical ganglion (in neck area)
   d) White rami communicantes and Gray rami communicantes
   e) Greater splanchnic nerve (labeled as Thoracic splanchnic nerve)

LABORATORY 13

Foramina of the Skull and Cranial Nerves

For the cranial nerves, you will need to know the following:

a) Nerve name and number (note that all cranial nerves are numbered using Roman numerals)
b) Nerve functions
c) The bony opening of the skull that each cranial nerve passes through to exit the cranial cavity
d) Important bony landmarks along the route of the nerve

Section I: ***AnatLab*** – Do the chapter for Cranial Nerves. You are not responsible for the mechanism by which visual information crosses over to the other side of the brain.

Section II: ***ADAM Interactive*** **– Foramen and Canals of the Skull**

Lateral View of the Skull

1. **File – Open – Content – Dissectible Anatomy**
2. Open a Lateral view of the head. Go to depth **layer 118** and identify the External auditory meatus.
3. At depth **layer 187**, identify the Mental foramen.
4. At depth **layer 208**, identify the Nasolacrimal canal.

Anterior View of the Skull

1. **File – Open – Content – Dissectible Anatomy**
2. Open an Anterior view of the head. Go to depth **layer 48** and identify the following:
   a) Supraorbital foramen
   b) Infraorbital foramen
   c) Mental foramen
   d) Optic canal
   e) Superior orbital fissure

Atlas Anatomy of the Skull

1. **File – Open – Content – Atlas Anatomy – System – Skeletal system**
2. Open **Skull (Inf)** and identify the following:
   a) Foramen lacerum
   b) Foramen ovale
   c) Stylomastoid foramen
   d) Foramen magnum
   e) Carotid canal (This is the entrance to the canal.)

1. **File – Open – Content – Atlas Anatomy – System – Skeletal system**
2. Open **Cranial Cavities (Sup) 2** and identify the following:
   a) Cribriform plate
   b) Optic canal
   c) Superior orbital fissure
   d) Foramen rotundum
   e) Foramen ovale
   f) Exit of Carotid canal – labeled as foramen lacerum
   g) Foramen spinosum
   h) Internal auditory meatus
   i) Jugular foramen
   j) Hypoglossal canal
   k) Foramen magnum

1. **File – Open – Content – Atlas Anatomy – System – Skeletal system**
2. Open **Walls of Orbit (Ant)** and identify the following:
   a) Optic canal
   b) Superior orbital fissure
   c) Nasolacrimal canal

Section III: *ADAM Interactive* – Cranial Nerves

Atlas Anatomy

File – Open – Content – Atlas Anatomy – System – Nervous system

1. *Olfactory Nerve (CN I).*
   a) Scroll down and select **Olfactory Nerve in Nasal Cavity**. Identify the following:
      1) Olfactory nerve – not a single nerve, but a series of small nerve fibers
      2) Cribriform plate of Ethmoid bone
      3) Superior nasal concha – Olfactory epithelium indicated in green
   b) Close window when finished
2. *Optic Nerve (CN II) and Extraocular Muscles.*
   a) In **Atlas Anatomy**, select **Optic Nerve in Orbit (Lat).** Identify the Optic nerve (Exits cranial cavity through what opening?)
   b) Close window when finished
3. *Oculomotor Nerve (CN III) and Extraocular Muscles.*
   a) In **Atlas Anatomy**, select **Oculomotor Nerve in Orbit (Lat).** Identify the following:
      1) Oculomotor nerve
      2) Ciliary ganglion
   b) Close window when finished
   c) In **Atlas Anatomy**, select **Oculomotor Nerve in Orbit (Ant).** Identify the following:
      1) Oculomotor nerve
      2) Ciliary ganglion
      3) Superior orbital fissure
   d) Close window when finished
4. *Trochlear nerve (CN IV).*
   a) In **Atlas Anatomy**, select **Trochlear Nerve in Orbit (Lat)**. Identify the following:
      1) Trochlear nerve
      2) Superior oblique muscle and Trochlea
   b) Close window when finished

5. *Trigeminal nerve (CN V).*
   a) In **Atlas Anatomy**, select **Dermatomes of Head/Neck (Lat)**. Identify the following:
      1) Dermatome of Ophthalmic division of Trigeminal nerve (V_1)
      2) Dermatome of Maxillary division of Trigeminal nerve (V_2)
      3) Dermatome of Mandibular division of Trigeminal nerve (V_3)
      4) Close window when finished
   b) In **Atlas Anatomy**, select **Trigeminal Nerve – V_1**. Identify the following:
      1) Ophthalmic division of Trigeminal nerve
      2) Superior orbital fissure
      3) Supraorbital nerve
      4) Supraorbital foramen
      5) Close window when finished
   c) In **Atlas Anatomy**, select **Trigeminal Nerve – V_2**. Identify the following:
      1) Maxillary division of Trigeminal nerve
      2) Foramen rotundum
      3) Infraorbital nerve
      4) Close window when finished
   d) In **Atlas Anatomy**, select **Trigeminal Nerve – V_3**. Identify the following:
      1) Mandibular division of Trigeminal nerve (Exits cranial cavity through what opening?)
      2) Inferior alveolar nerve
      3) Mental nerve
      4) Close window when finished
6. *Abducent Nerve (CN VI).*
   a) In **Atlas Anatomy**, select **Abducent Nerve in Orbit**. Identify the following:
      1) Abducent nerve
      2) Superior orbital fissure
   b) Close window when finished
7. *Facial Nerve (CN VII).*
   a) In **Atlas Anatomy**, select **Facial Nerve (Lat) 1**. Identify the following:
      1) Facial nerve

      This nerve exits the cranial cavity and enters the petrous portion of the temporal bone by passing through what opening? It then exits the base of the skull via what opening?

      2) Parotid salivary gland
   b) Close window when finished
8. *Vestibulocochlear Nerve (CN VIII).*
   a) In **Atlas Anatomy**, select **External, Middle, & Inner Ear**. Identify the following:
      1) Cochlear division of Vestibulocochlear nerve
      2) Vestibular division of Vestibulocochlear nerve
      3) Cochlea (Cochlear duct + Membranous labyrinth)
      4) Internal acoustic meatus
      5) External acoustic meatus (bony part)
   b) Close window when finished
9. *Glossopharyngeal Nerve (CN IX).*
   a) In **Atlas Anatomy**, select **Glossopharyngeal Nerve 1**. Identify the following:
      1) Glossopharyngeal nerve (Exits cranial cavity through what opening? exits base of skull through what opening?
      2) Posterior 1/3 of tongue (*not labeled*; area posterior to Terminal sulcus)

        3) Parotid gland
        4) Common carotid artery and Carotid body
    b) Close window when finished

10. *Vagus Nerve (CN X).*
    a) In **Atlas Anatomy**, select **Vagus Nerve (Lat)**. Identify the following:
        1) Vagus nerve
        2) Jugular foramen
        3) Vagal nerve branches (motor & sensory) to Pharynx and Larynx
        4) Cardiac branches (Preganglionic parasympathetic fibers to Heart & Lungs)
        5) Not shown – Preganglionic parasympathetic fibers to Abdominal viscera
    b) Close window when finished
11. *Accessory Nerve (CN XI).*
    a) In **Atlas Anatomy**, select **Accessory Nerve (Lat)**. Identify the following:
        1) Accessory nerve (Exits cranial cavity through what opening?)
        2) Sternocleidomastoid muscle
        3) Trapezius muscle
    b) Close window when finished
12. *Hypoglossal Nerve (CN XII).*
    a) In **Atlas Anatomy**, select **Ansa Cervicalis (Lat)**. Identify the Hypoglossal nerve.
    b) Close window when finished

LABORATORY 14

The Neck

Section I: *AnatLab* – Chapter for Head & Neck – Muscles & Vessels (concentrate on Neck structures)
Section II: *ADAM Interactive* – Part 1: Muscles of the Neck

Anterior View of Neck

1. Open **Dissectible Anatomy** – Anterior view of head and neck region.
2. Go to depth **layer 34** and identify the following:
   a) Sternocleidomastoid and Trapezius muscles. Define the triangles of the neck:
      1) *Anterior triangle*: Sternocleidomastoid muscle, Mandible, Anterior mid-line of neck
      2) *Posterior triangle*: Sternocleidomastoid muscle, Trapezius muscle, Clavicle
3. At depth **layer 53**, identify the layer of Superficial infrahyoid muscles:
   a) Superior and Inferior bellies of omohyoid muscle
   b) Sternohyoid muscle
4. Change to depth **layer 62** and identify the deeper layer of Infrahyoid muscles:
   a) Sternothyroid muscle
   b) Thyrohyoid muscle
   c) Thyroid cartilage of the larynx
   d) Hyoid bone
5. At depth **layer 78**, identify the Thyroid gland. Note that it has one lobe on each side of the Trachea. The lobes are connected to each other across the mid-line via a thin isthmus of thyroid tissue
6. At depth **layers 82-86**, identify the following:
   a) Anterior scalene muscle and Middle scalene muscle
   b) Note relationships of Subclavian artery and vein and Brachial plexus components to the Anterior scalene muscle
7. At depth **layers 277 & 328**, identify the Middle scalene muscle & Posterior scalene muscle, and Ribs 1-2.

Lateral View of the Neck

1. Use the **View** button to change to a Lateral view of the head and neck. Click **Normal Mode** button.
2. Go to depth **layer 8** and identify the following:
   a) Sternocleidomastoid muscle and Trapezius muscle; define the Anterior and posterior triangles of the neck

3. At depth **layers 41-43**, identify the following:
   a) Hyoid bone and Thyroid cartilage of the Larynx
   b) Superior and Inferior bellies of Omohyoid muscle
   c) Sternohyoid muscle, Sternothyroid muscle, and Thyrohyoid muscle
   d) Anterior and Posterior bellies of the Digastric muscle
4. At depth **layer 50**, identify the following:
   a) Mylohyoid muscle
   b) Stylohyoid muscle
5. At depth **layer 55**, identify the Anterior and Middle scalene muscles; note relationships of Subclavian artery and Trunks of Brachial plexus to these muscles. (What is the gap between these muscles called?)
6. At depth **layer 115**, identify the following:
   a) Thyroid gland
7. At depth **layer 209**, identify the following:
   a) Mylohyoid muscle (cut margin of muscle)

Atlas Anatomy – Muscles of the Neck

1. Review the muscles of head and neck region using Atlas Anatomy:
   File – Open – Content – Atlas Anatomy – Region – Head & Neck
2. Select **Hyoid Muscles (Lat)** and identify the following:
   a) Hyoid bone
   b) Anterior and Posterior bellies of Digastric muscle
   c) Stylohyoid muscle
   d) Mylohyoid muscle (cut margin of muscle)
   e) Superior and Inferior bellies of Omohyoid muscle; note fascial sling holding intermediate tendon to Clavicle
   f) Sternohyoid muscle
   g) Sternothyroid muscle (*not labeled)*
   h) Thyrohyoid muscle

1. **File – Open – Content – Atlas Anatomy – Region – Head & Neck**
2. Select **Triangles of the Neck (Ant)** and identify the following:
   a) Hyoid bone
   b) Clavicle
   c) Manubrium of sternum
   d) Sternocleidomastoid muscle
   e) Anterior and posterior bellies of digastric muscle; note fascia sling attaching to Hyoid bone
   f) Mylohyoid muscle
   g) Stylohyoid muscle
   h) Sternohyoid muscle
   i) Sternothyroid muscle
   j) Thyrohyoid muscle
   k) Anterior scalene muscle

Dissected cadaver images of the head and neck region are also found in Atlas Anatomy:
   a) **Region – Head & Neck** – chose from the following:
      1) **Posterior Triangle of Neck**
      2) **Superf. Muscles of Neck (Ant)**

(You are not responsible for being able to recognize structures from these cadaver images.)

Section III: ***ADAM Interactive* – Part 2: Blood Vessels and Nerves of the Head and Neck**

Anterior View of Head and Neck

1. Start with an Anterior view of head and neck region. Go to depth **layer 20** and identify the following:
   a) External jugular vein (Runs superficial to what muscle?)
2. At depth **layers 73-75**, identify the following:
   a) Carotid sheath (Contents?)
   b) Internal jugular veins, Subclavian veins, Axillary veins, and Brachiocephalic veins
   c) External carotid artery
3. At depth **layers 79-80**, identify the following:
   a) Common carotid artery, External carotid artery, and Internal carotid artery
   b) Vagus nerve and Phrenic nerve
4. At depth **layer 85**, identify the Internal thoracic artery. This branches off of the Subclavian artery, and then runs inferiorly into the Thoracic cavity. (Go to depth **layer 152** to see this artery within the Thorax).
5. Go to depth **layer 177**. In the upper thoracic region, identify the following:
   a) Arch of aorta
   b) Brachiocephalic artery dividing into Right common carotid artery and Right subclavian artery
   c) Left common carotid artery and Left subclavian artery – branch directly from the aorta
   d) Vagus nerve
6. At depth **layer 272**, identify the following:
   a) Vertebral artery
   b) C1-C4 spinal nerves (form Cervical plexus – *not labeled*)
   c) Phrenic nerve (Formed by what spinal nerve levels?)
   d) C5-T1 spinal nerves (Form Brachial plexus – *not labeled*)

Lateral View of the Neck

1. Use the **View** button to change to a Lateral view of the head and neck region. Click **Normal Mode** button.
2. Go to depth **layer 4** and identify the following:
   a) External jugular vein
3. At depth **layers 10-12**, identify the following:
   a) Sternocleidomastoid muscle
   b) Trapezius muscle
   c) Accessory nerve (CN XI) – passes deep to and innervates both muscles
   d) External jugular vein
4. At depth **layers 41-42**, identify the following:
   a) External carotid artery
   b) Internal jugular vein
   c) Subclavian vein
   d) Brachiocephalic vein
5. At depth **layer 49**, identify the following:
   a) Common carotid, External carotid, and Internal carotid arteries
   b) Vagus nerve (CN X)
   c) Cervical plexus (formed by C1-C4 spinal nerves)
   d) Phrenic nerve (formed by branches from C3-C5 spinal nerves)

6. At depth **layer 55**, identify the following structures, all of which pass through the Interscalene triangle (Borders of this triangle?):
   a) Trunks of Brachial plexus
   b) Branches of cervical plexus
   c) Subclavian artery
7. In the thoracic region at depth **layer 141**, identify the Internal thoracic artery. (Arises as a branch of what artery?)
8. At depth **layer 230**, identify the Vertebral artery (Branch of what artery? What artery does this become as it runs along the anterior side of the brainstem?)

LABORATORY 15

The Head

Section I: *AnatLab* – Chapter for Head & Neck – Muscles & Vessels (concentrate on Head structures)
Section II: *ADAM Interactive* – Part 1: Muscles of the Head and Neck

Anterior View of Head

1. Open **Dissectible Anatomy** – Anterior view of head and neck region.
2. Change depth to **layer 7** and identify the following:
   a) Platysma muscle
   b) Supraorbital nerve (Terminal branch of which trigeminal nerve division?)
   c) Infraorbital nerve (Terminal branch of which trigeminal nerve division?)
   d) Mental nerve (Terminal branch of which trigeminal nerve division?)
3. Go to depth **layer 11** and identify the following:
   a) Orbicularis oculi muscle
   b) Orbicularis oris muscle
   c) Parotid gland
   d) Branches of facial nerve (CN VII)
4. Go to depth **layer 34** and identify the following:
   a) Temporalis muscle
   b) Masseter muscle

Lateral View of the Head

1. Use the **View** button to change to a Lateral view of the head and neck. Click **Normal Mode** button.
2. At depth **layer 6**, identify the Platysma muscle.
3. Go to depth **layer 8** and identify the following:
   a) Orbicularis oculi muscle and Orbicularis oris muscle
   b) Parotid gland and Parotid duct, and Branches of Facial nerve (CN VII) emerging from the gland
4. At depth **layer 50**, identify the following:
   a) Buccinator muscle
   b) Masseter muscle
5. At depth **layer 115**, identify the following:
   a) Temporalis muscle (Name of bony fossa occupied by muscle?)
   b) Fibrous capsule of Temporomandibular joint

6. Go to depth **layers 122-124** and identify the following:
   a) Buccinator muscle
   b) Medial pterygoid muscle (Located within what bony fossa?)
   c) Lateral pterygoid muscle
7. At depth **layer 209**, identify the following:
   a) Submandibular gland and Sublingual gland

Atlas Anatomy – Muscles of the Head

1. **File – Open – Content – Atlas Anatomy – Region – Head & Neck**
2. Select **Temporomandibular Joint (Lat)** and identify the following:
   a) Condyle of mandible
   b) Mandibular fossa of Temporal bone
   c) Articular tubercle of Temporal bone
   d) Articular disc
   e) Lateral pterygoid muscle
   f) External auditory meatus
   g) Styloid process of Temporal bone

Dissected cadaver images of the head and neck region are also found in Atlas Anatomy:
   a) **Region – Head & Neck** – chose from the following:
      1) **Dissection of Facial Nerve**
      2) **Facial Nerve (Lat) 2**
      3) **Infratemporal Fossa 2**
      4) **Muscles of Facial Expression**
      5) **Orbicularis Oculi Muscle (Ant)**

(You are not responsible for being able to recognize structures from these cadaver images.)

Section III: *ADAM Interactive* – Part 2: Blood Vessels and Nerves of the Head and Neck

Anterior View of Head and Neck

1. Start with an Anterior view of head and neck region. Go to depth **layer 20** and identify the following:
   a) Facial artery and vein

Lateral View of the Head

1. Use the **View** button to change to a Lateral view of the head and neck region. Click **Normal Mode** button.
2. Go to depth **layer 4** and identify the following:
   a) Superficial temporal artery (A terminal branch of what artery?)
3. At depth **layers 41-42**, identify the following:
   a) Facial artery and vein
4. Change depth to **layers 121-123** and identify the following:
   a) Maxillary artery (A terminal branch of what artery?)
   b) Mandibular division of Trigeminal nerve (Exits skull via what opening?)
   c) Inferior alveolar nerve (seen entering the mandible) What is the terminal branch of this nerve?
5. At depth **layer 188**, identify the Middle meningeal artery as it supplies the Dura mater. (Branch of what artery? Passes through what opening in base of skull to enter cranial cavity?)

LABORATORY 16

Extraocular Muscles and Muscles of the Tongue

Section I: ***AnatLab*** – Do chapter for Cranial Nerves, concentrating on the eye and tongue muscles and their innervation

Section II: ***ADAM Interactive*** **–Extraocular Muscles**

Atlas Anatomy

File – Open – Content – Atlas Anatomy – System – Nervous system

1. *Optic Nerve (CN II) and Extraocular Muscles.*
   a) In **Atlas Anatomy**, select **Optic Nerve in Orbit (Lat)**. Identify the following:
      1) Optic nerve
      2) Levator palpebrae superioris muscle
      3) Superior, Inferior, Medial, and Lateral rectus muscles
      4) Superior oblique muscle and Trochlea
      5) Inferior oblique muscle
   b) Close window when finished
2. *Oculomotor Nerve (CN III) and Extraocular Muscles.*
   a) In **Atlas Anatomy**, select **Oculomotor Nerve in Orbit (Lat)**. Identify the following:
      1) Oculomotor nerve
      2) Ciliary ganglion
      3) Muscles innervated by Oculomotor nerve: Levator palpebrae superioris, Superior rectus, Medial rectus, Inferior rectus, and Inferior oblique muscles
   b) Close window when finished
   c) In **Atlas Anatomy**, select **Oculomotor Nerve in Orbit (Ant).** Identify:
      1) Oculomotor nerve
      2) Ciliary ganglion
      3) Muscles innervated by Oculomotor nerve: Levator palpebrae superioris, Superior rectus, Medial rectus, Inferior rectus, and Inferior oblique muscles
      4) Superior orbital fissure
   d) Close window when finished
3. *Trochlear nerve (CN IV).*
   a) In **Atlas Anatomy**, select **Trochlear Nerve in Orbit (Lat)**. Identify the following:
      1) Trochlear nerve
      2) Superior oblique muscle and Trochlea
   b) Close window when finished
4. *Abducent Nerve (CN VI).*
   a) In **Atlas Anatomy**, select **Abducent Nerve in Orbit**. Identify the following:
      1) Abducent nerve

2) Lateral rectus muscle
3) Superior orbital fissure

b) Close window when finished

Section III: *ADAM Interactive* – Muscles of the Tongue

Atlas Anatomy

1. In **Atlas Anatomy**, select **Vagus Nerve (Lat)**. Identify the following:
   a) Styloglossus muscle
   b) Hyoglossus muscle
   c) Genioglossus muscle

Dissectible Anatomy – Lateral View of the Tongue

1. Open **Dissectible Anatomy** – Lateral view of the head.
2. Go to depth **layer 233** and identify the following:
   a) Hypoglossal nerve (CN XII) and Lingual nerve (a branch of the Mandibular division of the Trigeminal nerve – V_3)
   b) Hyoglossus muscle
3. Go to depth **layers 242** and **245**. Identify the Genioglossus muscle. (Note that muscle arises from inner aspect of mandible and passes into entire tongue and also arches posteriorly to attach to the Hyoid bone.)

LABORATORY 17

Bones and Muscles of the Axial Skeleton

Section I: ***AnatLab*** – Do chapter for Axial Skeleton and Muscles (located within Upper Limb unit)
Section II: ***ADAM Interactive*** **– Bones of the Axial Skeleton and Muscles of the Deep Back**

Anterior View of Body

1. Open **Dissectible Anatomy** – Anterior view of chest region.
2. Change depth to **layer 56** and identify the following:
   a) Sternum:
      1) Manubrium, with Jugular notch, Clavicular notch, and Sternoclavicular joint
      2) Body
      3) Xiphoid process
   b) Ribs 1-11 (only tip of rib 11 seen; rib 12 not seen)
      1) Identify Costal cartilages
      2) Distinguish between True ribs, False ribs, and Floating ribs
   c) Identify the Sternal angle (*Not labeled.* Definition?)
3. Go to depth **layers 272 and 276-277** to identify the following:
   a) Cervical vertebrae:
      1) C1 vertebra (Atlas) – note that transverse process extends laterally to a greater degree than for other Cervical vertebrae. Identify the Atlanto-occipital joint (articulation between the Atlas and Occipital condyles of the Skull.) What motion is allowed at this joint?
      2) C2 vertebra (Axis) – go to depth **layer 330** to identify the Dens; note how it articulates with the inner wall of the Atlas. Identify the Atlantoaxial joint. What motion is allowed at this articulation?
      3) C3-C7 vertebrae; identify their Transverse processes and note how it is grooved to allow for passage of each spinal nerve
      4) Vertebral artery passing through the Transverse foramen of the Cervical vertebrae
   b) Thoracic vertebrae – T1-T12
      1) Indentify articulations between Head of rib and Body of Thoracic vertebrae
   c) Ribs 1-12; identify the Head and Neck of ribs
   d) Lumbar vertebrae; L1-L5, with Transverse processes
   e) Sacrum – formed by fusion of S1-S5 vertebraeanterior (ventral) sacral foramen – for exit of anterior branch of Sacral spinal nerves
   f) Intervertebral discs and anterior longitudinal ligament (uniting vertebral bodies)
4. At depth **layer 313**, identify the Coccyx.

Lateral View of Body

1. Open a Lateral view of the body. Go to depth **layers 253-255** and identify the following:
   a) Cervical vertebrae (C1-C7), Thoracic vertebrae (T-T12), and Lumbar vertebrae (L1-L5)
      1) Spinous processes (note that the C1 vertebra does not have a spinous process)
      2) Note U-shaped groove formed by Transverse process of Cervical vertebrae for passage of spinal nerve
   b) Sacrum and Coccyx
   c) Ribs – Head, Neck, and Angle
      1) Identify facet on body of Thoracic vertebra for articulation with Head of Rib
      2) Identify facet on Transverse process of Thoracic vertebra for articulation with Tubercle of rib
   d) Intervertebral discs
   e) Spinal nerves (emerging through Intervertebral foramen). Spinal nerves are seen in the thoracic and lumbar regions at depth **layer 243**. At deeper layers, the nerves are cut off at the point where they emerge from the Intervertebral foramen.
   f) Supraspinous ligament, and in the neck, the Ligamentum nuchae
   g) Note Cervical, Thoracic, Lumbar, and Sacrococcygeal curves of the vertebral column. Which are primary curves and which are secondary curves?

Posterior View of Body

1. Open a Posterior view of the body. Go to depth **layer 76** and identify the following:
   a) Ligamentum nuchae and Supraspinous ligament
   b) Erector spinae muscle columns:
      1) Iliocostalis muscle
      2) Longissimus muscle
      3) Spinalis muscle

(Note that each of these muscle columns are subdivided and given specific regional names. You are not responsible for these specific regional names.)

2. Go to depth **layer 176**. Identify the following:
   a) Cervical, Thoracic, and Lumbar vertebrae
      1) Spinous processes (Note that C1 vertebra (Atlas) lacks a spinous process.)
      2) Transverse processes
         (a) Note articulation of Rib Tubercle with Thoracic transverse processes
         (b) Note greater lateral extent of C1 vertebra compared to other Cervical vertebrae
      3) Lamina of vertebrae
      4) Superior and Inferior articular processes of vertebrae
   b) Sacrum and Coccyx
      1) Median sacral crest (Formed by fusion of what structures?)
      2) Lateral sacral crest (Formed by fusion of what structures?)
      3) Dorsal sacral foramen – for passage of posterior branch of Sacral spinal nerves
3. At depth **layer 180**, identify the following:
   a) Pedicle (cut) of vertebrae (Lamina and Spinous processes have been removed)
   b) Spinal nerves exiting vertebral column through Intervertebral foramen (*not labeled*)
4. Change depth to **layer 186** and identify the following:
   a) Dens of C2 vertebra (Axis)
   b) Bodies of C2-L5 vertebrae
   c) Tubercle of ribs
   d) Intervertebral discs

Atlas Anatomy

1. **File – Open – Atlas Anatomy – System – Skeletal – T12 Vertebrae (Sup)**
2. Identify the Intervertebral disc – Anulus fibrosus and Nucleus pulposus
3. Close window when finished.

1. **File – Open – Atlas Anatomy – System – Skeletal – Vertebral Column (Post)**
2. Identify the following:
   a) Cervical, Thoracic, and Lumbar vertebrae
   b) Spinous processes and Transverse processes of each vertebra
   c) Ribs 1-12
   d) Sacrum – Median sacral crest, Lateral sacral crest, Dorsal sacral foramen (*not labeled*)

1. **File – Open – Atlas Anatomy – System – Skeletal – Cervical Vertebrae (Ant/Lat)**
   File – Open – Atlas Anatomy – System – Skeletal – Cervical Vertebrae (Post/Lat)
2. Identify the following:
   a) C1 to C7 vertebra
   b) Dens (odontoid process) of C2
   c) Vertebral bodies
   d) Bifid spinous processes
   e) Transverse processes – note groove for passage of spinal nerve
   f) Intervertebral foramen

1. **File – Open – Atlas Anatomy – System – Skeletal – Cervical Spine (Odontoid)**
2. This is a radiograph of the Atlantoaxial joint, taken looking through the open mouth.
3. Identify the following:
   a) C1 (atlas) vertebra
   b) C2 (axis) vertebra with its Dens (odontoid process) and Transverse process
   c) Occipital condyle of Skull
   d) Atlantoaxial and Atlanto-occipital joints

1. **File – Open – Atlas Anatomy – System – Skeletal – Lumbar Vertebrae (Ant/Lat)**
   File – Open – Atlas Anatomy – System – Skeletal – Lumbar Vertebrae (Post/Lat)
2. Identify the following:
   a) L1 to L5 vertebrae
   b) Body and spinous process of each vertebrae
   c) Intervertebral foramen
   d) Note gaps between vertebrae for intervertebral discs

1. **File – Open – Atlas Anatomy – System – Skeletal – Lumbar Spine (Lat)**
2. This is a lateral radiograph (X-ray) of the lumbar vertebral column. Identify the following:
   a) Lumbar vertebrae – Body, Spinous process, Pedicle, Superior & Inferior articular processes, and Intervertebral foramen
   b) Note the large spaces between the bodies of the lumbar vertebrae. These spaces are occupied by the intervertebral discs, which are not dense enough to be seen in X-ray photographs.
3. Close window when finished.

LABORATORY 18

Bones of the Upper Limb

Section I: ***AnatLab*** – Do chapter for Bones of the Upper Limb
Section II: ***ADAM Interactive*** **– Bones of the Upper Limb**

Atlas Anatomy – Upper Limb Skeleton

1. **File – Open – Content – Atlas Anatomy – System – Skeletal**
2. Open **Bones of Arm & Shoulder (Ant)** and identify the following:
   a) Clavicle
   b) Sternum – Body, Manubrium
   c) Scapula – Acromion, Coracoid process
   d) Humerus – Head
3. **File – Open – Content – Atlas Anatomy – System – Skeletal**
4. Open **Bones of Arm & Shoulder (Post)** and identify the following:
   a) Scapula – Acromion, Spine
   b) Humerus
   c) Ulna – Olecranon process
5. **File – Open – Content – Atlas Anatomy – System – Skeletal**
6. Open **Bones of Forearm & Hand (Ant)** and identify the following:
   a) Radius – Head (proximal end of radius)
   b) Ulna – Head (distal end of ulna)
   c) Carpal bones (individual names of carpal bones given in Muscles of Forearm chapter)
   d) Metacarpal bones
   e) Phalanges (Phalanx bones)
   f) Interosseous membrane
7. **File – Open – Content – Atlas Anatomy – System – Skeletal**
8. Open **Bones of Forearm & Hand (Post)** and identify the following:
   a) Radius
   b) Ulna – Olecranon process
   c) Interosseous membrane

LABORATORY 19

Muscles of the Shoulder

Section I: ***AnatLab*** – Do chapter for Muscles of the Shoulder and Arm
Section II: ***ADAM Interactive*** **– Bones of the Shoulder Region**

Atlas Anatomy – Upper Limb Skeleton

1. **File – Open – Content – Atlas Anatomy – System – Skeletal**
2. Open **Bones of Arm & Shoulder (Ant)** and identify the following:
   a) Sternum – Body, Manubrium
   b) Clavicle
   c) Scapula
      1) Acromion
      2) Coracoid process
      3) Infraglenoid tubercle
   d) Humerus
      1) Head
      2) Greater tubercle - note that the greater tubercle is located laterally
      3) Lesser tubercle - note that the lesser tubercle is located anteriorly
      4) Medial epicondyle
      5) Lateral epicondyle
      6) Bicipital (intertubercular) groove
      7) Deltoid tuberosity
3. **File – Open – Content – Atlas Anatomy – System – Skeletal**
4. Open **Bones of Arm & Shoulder (Post)** and identify the following:
   a) Clavicle
   b) Scapula
      1) Acromion
      2) Spine
      3) Supraspinous fossa
      4) Infraspinous fossa
      5) Medial border
      6) Lateral border
      7) Superior angle
      8) Inferior angle
      9) Suprascapular (scapular) notch
      10) infraglenoid tubercle
   c) Humerus – Head and Greater tubercle

Dissectible Anatomy – Anterior View of Upper Limb Bones

1. Open **Dissectible Anatomy** – Start with an Anterior view of shoulder region.
2. Change depth to **layer 56** and identify the following:
   a) Clavicle – Sternal (medial) and Acromial (lateral) ends
   b) Scapula – Acromion process, Coracoid process (Note that the Coracoid process projects anteriorly just below the Clavicle.)
   c) Sternum – Clavicular notch and Jugular notch
   d) Sternoclavicular joint and Acromioclavicular joint
3. Go to depth **layers 328-329** and identify the following:
   a) Scapula – Articular cartilage, Lateral border
   b) Humerus – Articular cartilage, Head, Anatomical neck, Greater tubercle, Lesser tubercle, Bicipital (Intertubercular) groove, Deltoid tuberosity
   c) Glenohumeral joint (between Head of Humerus and Glenoid fossa of Scapula)

Lateral View of Pectoral Girdle

1. Use **View** button to change to Lateral view of the shoulder region. (Do not use Lateral Arm.) Click **Normal Mode** button.
2. Go to depth **layer 40** and identify the following in the Scapula:
   a) Acromion (Acromial process)
   b) Spine
   c) Coracoid process
   d) Glenoid cavity (Fossa)
   e) Supraspinous fossa
   f) Infraspinous fossa
   g) Medial border and Lateral border
   h) Superior angle and Inferior angle
   i) Infraglenoid tubercle

Posterior View of Upper Limb Bones

1. Use **View** button to change to Posterior view of the shoulder region. Click **Normal Mode** button.
2. Change depth to **layer 72** and identify the following on the scapula:
   a) Acromion (*cut*)
   b) Spine of scapula
   c) Supraspinous fossa
   d) Infraspinous fossa
   e) Medial border
   f) Lateral border
   g) Superior angle
   h) Inferior angle

Atlas Anatomy – Upper Limb Skeleton

1. **File – Open – Content – Atlas Anatomy – System – Skeletal**
2. Open **Shoulder (Ant/Post)**. In this radiograph (X-ray) of the shoulder region, identify the following:
   a) Clavicle

b) Scapula
   1) Acromion process – note that the Acromion process forms the superior tip of the Shoulder
   2) Coracoid process
   3) Glenoid cavity (Fossa)
   4) Supraglenoid tubercle
   5) Infraglenoid tubercle

c) Humerus
   1) Head
   2) Anatomical neck
   3) Surgical neck
   4) Greater tubercle
   5) Lesser tubercle

d) Glenohumeral joint

e) Acromioclavicular joint

Section III: *ADAM Interactive* – Muscles of the Shoulder

Posterior View of the Back and Shoulder Regions

1. Open **Dissectible Anatomy** – Start with a Posterior view of the shoulder and back regions.
2. Change depth to **layer 9** and identify the following:
   a) Trapezius muscle
   b) Latissimus dorsi muscle
   c) Deltoid muscle
3. Go to depth **layer 11** and identify the following:
   a) Levator scapulae muscle
   b) Rhomboid muscles (major and minor)
   c) Supraspinatus muscle (at depth **layer 18**, note how it passes under the Acromion process and over the Head of the Humerus)
   d) Infraspinatus muscle
   e) Teres minor muscle
   f) Teres major muscle
   g) Long head of Triceps brachii muscle
4. Click through depth **layers 19-29** to see the sequential removal of the Supraspinatus, Infraspinatus, Teres minor, and Long head of the Triceps brachii muscles. At depth **layer 29**, note that the Teres major muscle inserts onto the anterior side of the Humerus.
5. At depth **layer 38**, identify the Serratus Anterior muscle (Where on the scapula does it insert?)
6. At depth **layer 73**, identify the Subscapularis muscle. (Compare depth **layers 72** and **73**.)

Anterior View of the Chest and Arm Regions

1. Use the **View** button to select an Anterior view of the chest and shoulder regions. Click **Normal Mode** button.
2. Change depth to **layers 19-20** and identify the following:
   a) Deltoid muscle
   b) Pectoralis major muscle – Clavicular and Sternocostal heads. Note how the muscle has a half twist, with the highest arising muscle fibers inserting the most inferiorly on the Humerus.

3. Go to depth **layer 54** and identify the Pectoralis minor muscle (Insertion?)
4. At depth **layer 91**, identify the following:
   a) Serratus anterior muscle
   b) Long head of Triceps brachii muscle
   c) Latissimus dorsi muscle, Teres major muscle, and Subscapularis muscle inserting onto the anterior Humerus

Lateral View of the Chest and Arm Regions

1. Use the **View** button to select a <u>Lateral view of the chest and arm regions</u>. Click **Normal Mode** button.
2. Change depth to **layer 8** and identify the following:
   a) Trapezius muscle
   b) Pectoralis major muscle
   c) Deltoid muscle
   d) Latissimus dorsi muscle and Teres major muscle
   e) Rotator cuff muscles: Supraspinatus, Infraspinatus, Teres minor, and Subscapularis muscles
   f) Coracobrachialis muscle and Long head of Triceps brachii muscle
3. At depth **layer 13**, identify the following:
   a) Pectoralis minor muscle
   b) Infraspinatus muscle
   c) Teres minor muscle and Teres major muscle
   d) Serratus anterior muscle (Insertion? Compare depth **layers 41** and **45**.)
4. At depth **layers 25-26**, compare the areas of origin from the Scapula of the Teres major and Teres minor muscles.
5. At depth **layers 41-42**, identify the Subscapularis muscle.

LABORATORY 20

Muscles of the Arm

Section I: ***AnatLab*** – Do chapter for Muscles of the Shoulder and Arm.
Section II: ***ADAM Interactive*** **– Bones of the Arm, Forearm, and Elbow Regions**

Atlas Anatomy – Upper Limb Skeleton

1. **File – Open – Content – Atlas Anatomy – System – Skeletal**
2. Open **Bones of Arm & Shoulder (Ant)** and identify the following:
   a) Humerus
      1) Medial supracondylar ridge
      2) Lateral supracondylar ridge
      3) Medial epicondyle
      4) Lateral epicondyle
      5) Trochlea
      6) Capitulum
   b) Radius – Head, Radial tuberosity
   c) Ulna
3. **File – Open – Content – Atlas Anatomy – System – Skeletal**
4. Open **Bones of Arm & Shoulder (Post)** and identify the following:
   a) Humerus
      1) Radial groove
      2) Medial supracondylar ridge
      3) Lateral supracondylar ridge
      4) Medial epicondyle
      5) Lateral epicondyle
      6) Olecranon fossa
   b) Ulna – Olecranon process
   c) Radius
5. **File – Open – Content – Atlas Anatomy – System – Skeletal**
6. Open **Bones of Forearm & Hand (Ant)** and identify the followings:
   a) Humerus
      1) Medial supracondylar ridge
      2) Lateral supracondylar ridge
      3) Medial epicondyle
      4) Lateral epicondyle
      5) Trochlea
      6) Capitulum
   b) Radius – Head, Radial tuberosity
   c) Ulna – Coronoid process, Ulnar tuberosity

7. **File – Open – Content – Atlas Anatomy – System – Skeletal**
8. Open **Bones of Forearm & Hand (Post)** and identify the following:
   a) Humerus
      1) Medial supracondylar ridge
      2) Lateral supracondylar ridge
      3) Medial epicondyle
      4) Lateral epicondyle
      5) Olecranon fossa
   b) Ulna – Olecranon
   c) Radius – Head

Anterior View of Upper Limb Bones

1. Open **Dissectible Anatomy** – Start with an Anterior view of the elbow region.
2. At depth **layers 328-329**, identify the following:
   a) Humerus
      1) Medial and Lateral Supracondylar ridges
      2) Medial and Lateral Epicondyles
      3) Trochlea
      4) Capitulum
   b) Ulna – Coronoid process, Ulnar tuberosity
   c) Radius – Head, Radial tuberosity
   d) Articular cartilage of elbow joint
   e) Interosseous membrane of forearm

Posterior View of Upper Limb Bones

1. Use **View** button to change to Posterior view of the elbow region. Click **Normal Mode** button.
2. At depth **layer 72**, identify the following:
   a) Humerus – Medial and Lateral Epicondyles (*not labeled*)
   b) Ulna – Olecranon process (Note how this forms the tip of the elbow.)
   c) Radius – Head

Section III: *ADAM Interactive* – Muscles of the Arm

Posterior View of the Arm

1. Open **Dissectible Anatomy** – Start with a Posterior view of the arm region.
2. At depth **layer 9**, identify the following:
   a) Deltoid muscle
   b) Triceps brachii muscle – Long head and Lateral head
3. At depth **layer 28**, identify the Long and Lateral heads of the Triceps brachii muscle.
4. Go to depth **layer 29** and identify the Medial head of the Triceps brachii muscle. Note the nerve and artery running in the Radial groove of the Humerus, between the Medial and Lateral heads of the Triceps brachii muscle. (Compare to depth **layer 28**.)

Anterior View of the Arm

1. Use the **View** button to select an Anterior view of the shoulder and arm region. Click **Normal Mode** button.
2. Go to depth **layer 54** and identify the Biceps brachii muscle – Long head and Short head. (Origin of each?) Note that the Bicipital aponeurosis portion of the insertion for the Biceps brachii is incorrectly labeled as the Tendon of biceps brachii muscle.
3. At depth **layer 87**, identify the following:
   a) Coracobrachialis muscle
   b) Brachialis muscle (Go to depth **layers 125-126** to see the insertion for this muscle and its relationship to the Elbow joint.)
4. At depth **layer 91**, identify the following:
   a) Long head of Triceps brachii muscle
   b) Note how the tendon of the Long head of Biceps brachii muscle passes over the Head of the Humerus within the capsule of the Shoulder joint and emerges in the Bicipital (Intertubercular) groove of the Anterior humerus. (The Biceps brachii muscle and tendon are not shown, but the synovial tendon sheath is shown. This sheath surrounds the tendon and allows for friction-free movement of the tendon.)

Lateral View of the Shoulder

1. Use the **View** button to select a Lateral view of the body (not the Lateral Arm view). Position the screen for the shoulder area. Click **Normal Mode** button.
2. Change depth to **layer 8** and identify the following:
   a) Coracobrachialis muscle and Long head of Triceps brachii muscle
   b) Tendon of Long head of Biceps brachii muscle

L A B O R A T O R Y 21

Muscles of the Forearm

Section I: ***AnatLab*** – Do chapter for Muscles of the Forearm and Hand
Section II: ***ADAM Interactive*** **– Bones of the Forearm and Hand**

Atlas Anatomy – Bones of the Forearm and Hand

1. **File – Open – Content – Atlas Anatomy – System – Skeletal**
2. Open **Bones of Forearm & Hand (Ant)** or **Bones of Hand (Ant)**. Identify the following:
   a) Radius – Styloid process
   b) Ulna – Head, Styloid process
   c) Carpal bones (*"Tiny toads can hop, so let them play"*)
      1) Starting at base of thumb, distal row, from lateral to medial: Trapezium, Trapezoid, Capitate, Hamate
      2) Starting at base of thumb, proximal row, from lateral to medial: Scaphoid, Lunate, Triquetrum, Pisiform
   d) Metacarpal bones 1-5
   e) Phalanges (Phalanx bones)
      1) Proximal and Distal Phalanx bones for the thumb (digit 1)
      2) Proximal, Middle, and Distal Phalanx bones for the fingers (digits 2-5)
3. **File – Open – Content – Atlas Anatomy – System – Skeletal**
4. Open **Bones of Forearm & Hand (Post)** or **Bones of Hand (Post)**. Identify the following:
   a) Radius – Styloid process, Dorsal tubercle
   b) Ulna – Head, Styloid process
   c) Carpal bones
   d) Metacarpal bones 1-5
   e) Phalanges (Phalanx bones)
      1) Proximal and Distal Phalanx bones for the thumb (digit 1)
      2) Proximal, Middle, and Distal Phalanx bones for the fingers (digits 2-5)

Dissectible Anatomy – Anterior View of Upper Limb Bones

1. Open **Dissectible Anatomy** – Start with an Anterior view of hand and wrist region.
2. At depth **layer 329**, identify the following:
   a) Ulna – Styloid process
   b) Radius – Head, Styloid process
   c) Carpal bones (*"Tiny toads can hop, so let them play"*)
      1) Starting at base of thumb, distal row, from lateral to medial: Trapezium, Trapezoid, Capitate, Hamate

    2) Starting at base of thumb, proximal row, from lateral to medial: Scaphoid, Lunate, Triquetrum, Pisiform
  d) Metacarpal bones 1-5
  e) Phalanges (Phalanx bones)
    1) Proximal and Distal Phalanx bones for the thumb (digit 1)
    2) Proximal, Middle, and Distal Phalanx bones for the fingers (digits 2-5)
3. At depth **layer 330**, identify the following:
  a) Radiocarpal joint (*not labeled*)
  b) Articular disc of distal Radioulnar joint. Note that the Distal ulna bone does not articulate with the Carpal bones. This gap is filled by the Articular disc.

Posterior View of Upper Limb Bones

1. Use **View** button to change to Posterior view of the hand and wrist region. Click **Normal Mode** button.
2. At depth **layer 72**, identify the following:
  a) Radius – Styloid process, Dorsal tubercle (bony bump between grooves for Extensor pollicis longus muscle and Extensor carpi radialis brevis muscle)
  b) Ulna – Head, Styloid process

Section III: *ADAM Interactive* – Anterior Muscles of the Forearm

Anterior View of Forearm

1. Open **Dissectible Anatomy** – Start with an Anterior view of forearm and hand.
2. At depth **layer 9**, identify the Antebrachial fascia of the Forearm (labeled only as Fascia).
3. Change depth to **layer 81** and identify the following:
  a) Flexor retinaculum
  b) Palmar aponeurosis
  c) Superficial muscles of the Flexor forearm:
    1) Pronator teres muscle
    2) Flexor carpi radialis muscle (Go to depth **layer 139** to see its insertion.)
    3) Palmaris longus muscle (Note insertion into Palmar aponeurosis.)
    4) Flexor digitorum superficialis muscle
    5) Flexor carpi ulnaris muscle
4. Go to depth **layer 116** and identify the Flexor digitorum superficialis muscle. Follow its tendons into the hand. Note how the tendons split just prior to inserting onto the medial-lateral sides of the middle phalanx bone of digits 2-5. Go to depth **layer 117** to see how this split allows the tendons of the Flexor digitorum profundus muscle to continue down the fingers to insert on the Distal phalanx bone of these digits.
5. Reposition the screen to show the forearm region and go to depth **layer 129**. Identify the deeper muscles of the Flexor forearm:
  a) Flexor digitorum profundus muscle
  b) Flexor pollicis longus muscle
  c) Pronator quadratus muscle (Better seen at depth **layer 132**.)

Section IV: *ADAM Interactive* – Posterior Muscles of the Forearm

Posterior View of Forearm

1. Use the **View** button to change to a Posterior view of the forearm and hand. Click **Normal Mode** button.
2. Go to depth **layer 23** and identify the Extensor retinaculum.
3. At depth **layer 29**, identify the superficial muscles of the extensor forearm:
   a) Brachioradialis muscle (mostly on anterior side of forearm, with little visible in this posterior view)
   b) Extensor carpi radialis longus muscle (Go to depth **layers 58 and 62** to see its insertion.)
   c) Extensor carpi radialis brevis muscle (Go to depth **layers 58 and 62** to see its insertion.)
   d) Extensor digitorum muscle; follow its tendons into the hand, noting how it splits into 4 interconnected tendons and then inserts into the Extensor Expansion of digits 2-5
   e) Extensor digiti minimi muscle
   f) Extensor carpi ulnaris muscle
   g) Extensor expansion of digits 2-5 (Which muscles insert into these?)
4. Change the depth to **layer 58** to identify the deep muscles of the extensor compartment:
   a) Supinator muscle (Best seen at **layer 49**.)
   b) Abductor pollicis muscle
   c) Extensor pollicis brevis muscle
   d) Extensor pollicis longus muscle
   e) Extensor indicis muscle
   f) Note how the tendons of Abductor pollicis, Extensor pollicis brevis, and Extensor pollicis longus muscles pass laterally into the thumb and, in doing so, cross over (overlie) the tendons of Extensor carpi radialis longus and Extensor carpi radialis brevis muscles. Go back to depth **layer 40** to see these thumb muscles as they emerge from deep within the posterior forearm and cross over the Extensor carpi radialis longus and Extensor carpi radialis brevis muscles. This is why the Abductor pollicis, Extensor pollicis brevis, and Extensor pollicis longus muscles are called "outcropping" muscles.

Posterior View of Hand

1. Reposition the screen to a Posterior view of the hand.
2. Go to depth **layer 23** and identify the Extensor retinaculum.
3. Change depth to **layer 42** and identify the Extensor expansion covering the posterior sides of digits 2-5.

Anterior View of Posterior Forearm Muscles

1. Use the **View** button to change to a Anterior view of the forearm and hand. Click **Normal Mode** button.
2. Go to depth **layer 81** and identify the Brachioradialis muscle. (Action?)
3. Go to depth **layer 134** and identify the Supinator muscle.

Atlas Anatomy

1. **File – Open – Content – Atlas Anatomy – Region – Upper Limb**
2. Open **Flexor Tendons of the Hand** and identify the following:
   a) Flexor digitorum superficialis (Note bifurcation of tendon prior to insertion.)
   b) Flexor digitorum profundus (Note relation to tendons of Flexor digitorum superficialis.)
   c) Flexor pollicis longus

1. **File – Open – Content – Atlas Anatomy – Region – Upper Limb**
2. Open **Extensor Tendons of the Hand** and identify the following:
   a) Extensor digiti minimi
   b) Extensor digitorum (Note tendinous interconnections between tendons to digits 2-5.)
   c) Extensor indicis
   d) Extensor pollicis longus
   e) Extensor pollicis brevis
   f) Abductor pollicis longus
   g) Extensor expansion of digits 2-5
   h) Extensor carpi radialis longus
   i) Extensor carpi radialis brevis

LABORATORY 22

Muscles of the Hand

Section I: ***AnatLab*** – Do chapter for Muscles of the Forearm and Hand
Section II: ***ADAM Interactive*** **– Intrinsic Muscles of the Hand**

Dissectible Anatomy – Anterior View of Hand

1. Open **Dissectible Anatomy**. Start with an Anterior view of the hand.
2. Go to depth **layer 83** and identify the following:
   a) Palmar aponeurosis (What muscle inserts into this layer?)
   b) Flexor retinaculum (superficial part)
3. At depth **layer 101**, identify the following:
   a) Flexor retinaculum (deep part) that forms the roof of the Carpal tunnel
   b) Abductor pollicis brevis muscle
   c) Flexor pollicis brevis muscle
   d) Abductor digiti minimi muscle
   e) Flexor digiti minimi muscle
4. At depth **layers 111-113**, identify the following:
   a) Opponens pollicis muscle
   b) Opponens digiti minimi muscle
5. Change the depth to **layer 129** and identify the Lumbrical muscles. (Origin associated with tendons of what muscle? Lumbrical insertion? Lumbrical actions?)
6. At depth **layer 132**, identify the Adductor pollicis muscle (triangular-shaped muscle that arises via 2 heads; you are not responsible for knowing names of individual heads). This muscle is located on the anterior (palmar) side of the thumb web space (space between 1st and 2nd metacarpal bones).
7. At depth **layer 325**, identify the Palmar Interosseous muscles. (Action?)
8. Go to depth **layer 326** and identify the Dorsal interosseous muscles. (Action?) Note that the 1st dorsal muscle is located on the posterior side of the thumb web space. The 1st dorsal interosseous muscle acts on the index finger (digit 2). What action does this muscle produce?

Posterior View of Hand

1. Use the **View** button to change to a Posterior view of the hand. Click **Normal Mode** button.
2. Change depth to **layer 42** and identify the Extensor expansion covering the posterior sides of digits 2-5.
3. At depth **layer 67**, identify the Dorsal Interosseous muscles.

Atlas Anatomy – Anterior Hand

1. **File – Open – Content – Atlas Anatomy – Region – Upper Limb**
2. Open **Superficial Muscles of Palm** and identify the following:
   a) Abductor digiti minimi
   b) Flexor digiti minimi
   c) Opponens digiti minimi
   d) Lumbrical muscles
   e) Abductor pollicis brevis
   f) Flexor pollicis brevis
   g) Opponens pollicis
   h) Adductor pollicis
   i) Flexor retinaculum
   j) Tendons of flexor digitorum superficialis
3. **File – Open – Content – Atlas Anatomy – Region – Upper Limb**
4. Open **Flexor Tendons of the Hand** and identify the following:
   a) Abductor digiti minimi
   b) Opponens pollicis
   c) Adductor pollicis
   d) Flexor digitorum superficialis
   e) Flexor digitorum profundus
   f) Lumbrical muscles (Origin? Insertion? Actions?)
   g) Flexor pollicis longus

Atlas Anatomy – Posterior Hand

1. **File – Open – Content – Atlas Anatomy – Region – Upper Limb**
2. Open **Extensor Tendons of the Hand** and identify the First dorsal interosseous muscle – forms posterior (dorsal) side of thumb web space.

LABORATORY 23

Nerves and Blood Vessels of the Upper Limb

Section I: ***AnatLab*** – Do upper limb chapter for Nerves, Blood Vessels
Section II: ***ADAM Interactive*** **– Nerves and Blood Vessels of the Upper Limb**

Anterior View of Shoulder Region

1. Open **Dissectible Anatomy** – Start with an Anterior view of the shoulder region.
2. At depth **layer 5**, identify the Cephalic vein as it dives through the Deltopectoral triangle to drain into the Axillary vein (junction seen in depth **layer 53**).
3. At depth **layer 25** (immediately deep to the Pectoralis major muscle), identify the following:
   a) Lateral pectoral nerve – innervates Pectoralis major muscle
   b) Medial pectoral nerve – innervates both Pectoralis major and Pectoralis minor muscles. Note how the nerve passes through the Pectoralis minor muscle.

   NOTE: Medial and lateral pectoral nerves are not named for their locations on the anterior chest wall (locations are actually the opposite of their names). They are named for the brachial plexus cord from which they are derived.
4. Change depth to **layer 85** and identify the following:
   a) Subclavian artery, Axillary artery, and Brachial artery (What anatomical landmarks serve as dividing lines between each of these?)
   b) Medial cord and Lateral cord of Brachial plexus (these are named for their locations, medial and lateral to what?)
   c) Medial pectoral nerve and Lateral pectoral nerve (from Medial and Lateral cords, respectively)
   d) Long thoracic nerve to Serratus anterior muscle
   e) Musculocutaneous nerve to all muscles of anterior arm. This nerve is the direct continuation of the Lateral cord. Distinguishing feature: Musculocutaneous nerve passes through the Coracobrachialis muscle.
   f) Median nerve, with its Lateral root from the Lateral cord and Medial root from the Medial cord. The Median nerve runs in the Neurovascular compartment down the Medial arm, then crosses the elbow region by passing through the Cubital fossa.
   g) Ulnar nerve – This is the direct continuation of the Medial cord. It passes down the Medial arm.

   NOTE: A distinguishing feature for the Musculocutaneous nerve, Median nerve (with its medial and lateral roots), and the Ulnar nerve is the capital "M" shape these nerves form as they emerge from the Brachial plexus.

5. At depth **layer 88**, identify the following:
   a) Branch of Subclavian artery – Dorsal scapular artery (passing to posterior shoulder region)
   b) Branches of Axillary artery:
      1) Lateral thoracic artery – runs with Long thoracic nerve along Lateral thoracic wall
      2) Subscapular artery – quickly divides into the following:
         (a) Thoracodorsal artery – with Thoracodorsal nerve to supply Latissimus dorsi muscle
         (b) Circumflex Scapular artery – through Triangular space (borders?) to posterior shoulder region
      3) Anterior and Posterior Circumflex Humeral arteries – wrap around Proximal humerus, deep to deltoid muscle
   c) Branch of Brachial artery – Deep Brachial artery (passes posterior to Humerus, running between Medial and Lateral heads of Triceps brachii muscle, accompanied by Radial nerve)
6. Change the depth to **layer 90** and identify the following:
   a) Posterior cord of brachial plexus
   b) Axillary nerve – passes through Quadrangular space (borders?) to posterior shoulder region
   c) Radial nerve – passes posterior to Humerus along Radial (spiral) groove of Humerus
   d) Long thoracic nerve – to Serratus anterior muscle
   e) Thoracodorsal nerve – to Latissimus dorsi muscle
   f) Subscapular nerves (upper and lower branches) – to Subscapularis and Teres major muscles
   g) Suprascapular nerve – to Supraspinatus and Infraspinatus muscles
7. Go to depth **layer 173 & 177** and reposition image to view the upper chest and base of neck regions.
   Identify the following:
   a) Arch of Aorta
   b) Brachiocephalic trunk – divides into Right Subclavian artery and Right Common Carotid artery
   c) Left Common Carotid artery and Left Subclavian artery
8. At depth **layer 274**, identify the following:
   a) C5 – T1 Spinal nerves that form the Brachial plexus
   b) Superior, Middle, and Inferior trunks of the Brachial plexus
   c) Anterior and Posterior divisions of each trunk
   d) Medial, Lateral, and Posterior cords of Brachial plexus

Anterior View of Cubital Fossa and Forearm Regions

1. Reposition screen to the Anterior forearm and elbow regions. Go to depth **layer 5**. Identify these subcutaneous veins (located superficial to the Brachial and Antebrachial fascia layers):
   a) Cephalic vein – passes up Lateral forearm and arm
   b) Basilic vein – passes up Medial forearm and arm; dives deep (penetrates Brachial fascia) in mid-arm region.
   c) Median cubital vein – overlies Cubital fossa region; connects Cephalic vein to Basilic vein (this vein is seen in *ADAM* software, but figure shown here has a different pattern of veins)

2. Change depth to **layer 84** and identify the following:
   a) Median nerve and Brachial artery – both pass through Cubital fossa (borders?), anterior to Elbow joint
   b) Ulnar nerve – disappears as it passes posterior to Medial epicondyle of Humerus
3. At depth **layer 103**, identify the following:
   a) Radial nerve – crosses Elbow joint by passing anterior to Lateral epicondyle of Humerus
   b) Terminal branches of Brachial artery – Radial artery (runs down lateral side of forearm; follow it to Lateral wrist) and Ulnar artery (runs down medial side of forearm)
4. At depth **layer 118**, identify the following:
   a) Median nerve in Cubital fossa region; runs down forearm between Flexor digitorum superficialis and Flexor digitorum profundus muscles (What muscles does it innervate in the forearm?)
   b) Ulnar nerve – seen emerging from Posterior to Medial epicondyle of Humerus; continues down Medial forearm (Muscles innervated in forearm?)
   c) Ulnar artery – passes to down forearm to Medial wrist
5. Go to depth **layer 123**. Identify the Common Interosseous artery as a branch from the Ulnar artery. The Common interosseous artery divides to become the Anterior interosseous artery and the Posterior Interosseous artery (posterior branch not seen here; anterior branch better seen at depth **layer 140**)

Anterior View of the Wrist and Hand Regions

1. Reposition the screen to the <u>Anterior wrist and hand regions</u>. Go to depth **layer 99**. Identify the following:
   a) Radial artery (at Lateral wrist) and Ulnar artery (at Medial wrist)
   b) Superficial palmar arterial arch – receives major blood supply from Ulnar artery
   c) Median nerve – passes deep to Flexor retinaculum within Carpal tunnel; Recurrent median nerve – motor branch to Thenar compartment muscles (Names of muscles?)
   d) Ulnar nerve – passes superficial to Flexor retinaculum (Muscles innervated in hand?)
2. Go to depth **layer 137** and identify the following:
   a) Deep palmar arterial arch – receives major blood supply from Radial artery (seen emerging into Deep palm from between 1st and 2nd Metacarpal bones)
   b) Deep branches of Ulnar nerve to Interosseous muscles

Posterior View of Shoulder Region

1. Use the **View** button to change to a <u>Posterior view of the shoulder region</u>. Click **Normal Mode** button.
2. At depth **layer 20**, identify the following:
   a) Circumflex Scapular artery as it emerges through the Triangular Space (Borders?)
   b) Posterior circumflex humeral artery and Axillary nerve as they emerge through the Quadrangular Space (Borders?)
3. Go to depth **layer 35** and identify the following:
   a) Dorsal Scapular artery and nerve – run along Medial border of Scapula (deep to and supplying Levator scapulae and Rhomboid muscles)
   b) Suprascapular artery and nerve – supply Supraspinatus muscle (in Supraspinous fossa) and Infraspinatus muscle (in Infraspinous fossa)
   c) Subscapular artery – divides into the following:

1) Thoracodorsal artery – runs with Thoracodorsal nerve deep to (and supplies) Latissimus dorsi muscle
2) Circumflex scapular artery – supplies Infraspinatus muscle

d) Posterior Circumflex Humeral artery and Axillary nerve – wrap around Posterior humerus, deep to Deltoid muscle

4. Reposition screen to view Posterior arm and elbow regions. At depth **layer 35**, identify the following:
   a) Radial nerve and Deep Brachial artery running posterior to Humerus in Radial groove. (Deep brachial artery better seen at depth **layer 57**.) Note that the Radial nerve crosses the Elbow by passing anterior to the Lateral epicondyle of the Humerus.
   b) Ulnar nerve – seen crossing Elbow posterior to Medial epicondyle of Humerus

Posterior View of Hand and Wrist Regions

1. Reposition screen to view Posterior wrist and Hand regions. Go to depth **layer 3** and identify the following:
   a) Dorsal venous arch – complex series of interconnected veins that receives venous blood from the fingers
   b) Cephalic vein – arises from lateral side of Dorsal arch and passes to Lateral forearm
   c) Basilic vein – arises from medial side of Dorsal arch and passes to Medial forearm
2. At depth **layer 50**, identify the following:
   a) Radial nerve
   b) Posterior interosseous artery
3. At depth **layer 65**, identify the following:
   a) Radial artery – crosses Lateral side of wrist, then dives through first Dorsal interosseous muscle (passes between 1st and 2nd Metacarpal bones) to supply Deep palmar arterial arch

Atlas Anatomy

1. **File – Open – Content – Atlas Anatomy – Region – Upper Limb**
2. Open **Anatomical Snuff Box**. In this cadaver view of the Lateral hand, identify the following:
   a) Tendons of Abductor pollicis and Extensor pollicis brevis (form lateral border of Snuff box)
   b) Tendon of Extensor pollicis longus (forms medial border of Snuff box)
   c) Radial artery (passes through Snuff box)
   d) Tendon of Extensor carpi radialis longus
   e) Tendon of Extensor carpi radialis brevis
   f) First Dorsal interosseous muscle (forms posterior side of thumb web space)

Use the following Atlas Anatomy figures to identify the arteries and nerves listed in your course pack lecture notes:

File – Open – Content – Atlas Anatomy – Region – Upper Limb – Arteries of Upper Limb (Ant)
File – Open – Content – Atlas Anatomy – Region – Upper Limb – Arteries of Upper Limb (Post)
File – Open – Content – Atlas Anatomy – Region – Upper Limb – Nerves of Upper Limb (Ant)
File – Open – Content – Atlas Anatomy – Region – Upper Limb – Nerves of Upper Limb (Post)

LABORATORY 24

Joints of the Upper Limb

Section I: ***AnatLab*** – Do Review and Quiz chapter for Upper Limb
Section II: ***ADAM Interactive* – Joints of the Upper Limb**

Anterior View of Shoulder Joint

1. Open an Anterior view of the shoulder region. Go depth **layer 53** and identify the following:
   a) Sternoclavicular joint and Acromioclavicular joint (joints are covered by ligaments)
   b) Coracoclavicular ligament – strongest connection between Scapula and Clavicle
2. At depth **layer 81**, identify the Tendon of Long head of Biceps brachii muscle as it emerges from within the Articular capsule of the Shoulder joint.
3. At depth **layer 95**, note how the tendons of the Supraspinatus and Subscapularis muscles fuse into the fibrous Articular capsule of the Shoulder (glenohumeral) joint
4. Change to depth **layer 145** and identify the following:
   a) Articular cartilages of Humerus and Scapula
   b) Layers of joint capsule – outer Fibrous capsule and inner lining of Synovial capsule (membrane)
5. At depth **layer 330**, identify the Glenoid Labrum at the margins of the Glenoid fossa of the Scapula.

Lateral View of the Shoulder Joint

1. Use the **View** button to change to a Lateral view of the shoulder region. Click the **Normal Mode** button.
2. At depth **level 0**, identify the following:
   a) Glenoid cavity (fossa) of Scapula
   b) Glenoid labrum
   c) Synovial capsule (membrane) of Shoulder joint
   d) Tendon of Long head of Biceps brachii muscle –within Articular capsule as it passes over Head of Humerus
   e) Rotator Cuff (*not labeled*) – formed by fusion with the fibrous Articular capsule of the tendons from the Subscapularis, Supraspinatus, Infraspinatus, and Teres minor muscles

LABORATORY 25

Muscles of Abdominal and Pelvic Walls

Section I: ***AnatLab*** – Do Abdomen–Thorax chapter for Muscles of Abdomen & Digestive System
Section II: ***ADAM Interactive*** **– Muscles of the Abdominal and Pelvic Walls**

Anterior View of Male Abdomen

1. Open **Dissectible Anatomy** – Start with Male – Anterior view of the abdomen and pelvic regions.
2. Go to depth **layer 18** and identify the following:
   a) External abdominal oblique muscle and Aponeurosis
   b) Linea alba
   c) Inguinal ligament (Medial and lateral bony attachment points?)
   d) Superficial inguinal ring
   e) Spermatic cord (*not labeled*) exiting from Superficial inguinal ring – Spermatic cord seen here covered by its outermost layer (External spermatic fascia)
3. At depth **layer 26**, identify the Internal abdominal oblique muscle and Aponeurosis, and Linea alba.
4. At depth **layer 29**, identify the following:
   a) Transversus abdominis muscle
   b) Rectus abdominis muscle, with Tendinous intersections
5. Change the depth to **layer 55** and identify the Posterior layer of Rectus sheath.
6. At depth **layer 157**, identify the Deep inguinal ring.
7. At depth **layers 178-180**, identify the following:
   a) Testicular vein (labeled as Pampiniform venous plexus)
   b) Testicular artery
   c) Ductus deferens
   d) Testis
   e) Note attachments of Inguinal ligament to Anterior superior iliac spine and Pubic tubercle (*bony landmarks are not labeled*)
8. Change depth to **layer 266** and identify the following:
   a) Psoas major muscle
   b) Iliacus muscle
9. At depth **layer 278**, identify the Quadratus lumborum muscle.

Lateral View of Abdomen

1. Use the **View** button to change to a Lateral view of the Abdomen. Click **Normal Mode** button.
2. Go to depth **layers 11 and 12**. Identify the External abdominal oblique muscle and its Aponeurosis.
   Note orientation (direction) of muscle fibers.
3. Go to depth **layer 61** and identify the following:
   a) Internal abdominal oblique muscle and Aponeurosis (Note orientation of muscle fibers.)
   b) Thoracolumbar aponeurosis (*Not labeled* – labeled as Fascia.)
4. Go to depth **layer 67** and identify the Rectus abdominis muscle.
5. Go to depth **layer 72** and identify the following:
   a) Transversus abdominis muscle and Aponeurosis (Note muscle fiber orientation.)
   b) Thoracolumbar aponeurosis (*not labeled*)

Medial View of Male and Female Pelvis

1. Use the **View** button to change to a Medial view of the Male (or Female) pelvic region. Click **Normal Mode** button.
2. Go to depth **layer 49** and identify the following:
   a) Rectum and Anus (both sexes)
   b) Urethra (both sexes)
   c) Vagina (female)
3. Go to depth **layer 52** (**layer 56** in female) and identify the following:
   a) Pelvic Diaphragm (*not labeled*; identified as Coccygeus and Levator ani muscles; you are not required to know these muscle names)
   b) Note that Pelvic diaphragm arises along U-shaped line around inner margins of Lesser (true) pelvic cavity and then tapers to narrow inferior opening that allows for passage of Rectum. Anterior wall of pelvic diaphragm has a gap (Urogenital hiatus – not labeled; represents top of U-shape). Passing through the gap of the urogenital hiatus is the Urethra (males and females) and the Vagina (females)
4. Use the **Gender** button to change sex of image and repeat this section. Click **Normal Mode** button.

Atlas Anatomy

1. **File – Open – Content – Atlas Anatomy – System – Digestive – Pelvic Diaphragm (Med)**
2. This shows the Pelvic diaphragm in isolation. Identify the following:
   a) Pelvic diaphragm (Labeled as Coccygeus and Levator ani muscles; you are not responsible for these names.)
   b) Note that the Pelvic diaphragm is a funnel-shaped structure, with its larger, superior opening arising from the inner walls of the Lesser pelvic cavity and its inferior, smaller opening allowing for passage of the Rectum. (Rectum not seen in this view.)
   c) Anus and External anal sphincter muscle

1. **File – Open – Content – Atlas Anatomy – System – Digestive – Female Pelvic Diaphragm (Inf)**
2. Identify the following:
   a) Muscle of Pelvic diaphragm (Labeled as Levator ani and Coccygeus muscles; you are not responsible for knowing these namesL)
   b) Genital (urogenital) hiatus
   c) Urethra
   d) Vagina

LABORATORY 26

Organs of the Digestive System

Section I: ***AnatLab*** – Do Abdomen–Thorax chapter for Muscles of Abdomen & Digestive System
Section II: ***ADAM Interactive*** **– Organs of the Digestive Tract**

Anterior View of Abdominal Cavity

1. Use the **View** button to open an Anterior view of the abdominal region. Click **Normal Mode** button.
2. Change to depth **layer 198** and identify the following:
   a) Liver (right and left lobes)
   b) Gall bladder – seen at inferior margin of right lobe of Liver; compare to depth **layer 199**
   c) Stomach
   d) Cecum
   e) Ascending, Transverse and Sigmoid Colon
   f) Jejunum and Ileum of Small Intestine
   g) Compare depth **layers 158** and **198** – consider locational relationships between Rib cage and the Liver-Stomach. Note that both organs are located superior to the Inferior margin of the Rib cage. Also note the relationship between the Intestines and the Iliac crest. The Cecum, Sigmoid colon, and much of the Small intestine are located below the level of the Iliac crest (within the Greater pelvic cavity).
3. At depth **layer 202**, identify the following:
   a) Esophagus
   b) Stomach
      1) Greater curvature
      2) Lesser curvature
      3) Pyloric sphincter region (labeled as Pylorus)
      4) Esophageal sphincter region (*not labeled*)
   c) Duodenum (only first part seen at this level)
   d) Gall bladder
   e) Cystic duct and Common Hepatic duct
4. At depth **layer 205**, identify the following in the stomach:
   a) Rugae
   b) Pyloric sphincter
5. At depth **layer 206**, identify the following:
   a) Large intestine:
      1) Teniae coli
      2) Epiploic appendages

3) Haustra coli (*not labeled*)
4) Ascending colon, Transverse colon
5) Cecum
6) Hepatic (right) flexure (Separates which 2 parts of large intestine?)
7) Splenic (left) flexure (Separates which 2 parts of large intestine?)

6. At depth **layer 209**, identify the following:
   a) Cecum and Appendix
   b) Ascending, Descending, and Sigmoid Colon
   c) Rectum
   d) Ileocecal junction
7. Change depth to **layer 214**. Identify the following:
   a) Duodenum
   b) Pancreas – with Head and Tail
   c) Spleen
8. At depth **layer 215**, identify the Pancreatic Duct and Common Bile Duct. (Note how these join together at the point where they empty into the Duodenum.)
9. At depth **layer 221**, identify the Common Hepatic duct and the Cystic duct, joining together to form the Common Bile duct.

Atlas Anatomy

1. Open **Atlas Anatomy – System – Digestive – Abdominopelvic Quadrants.** This shows the abdominal organs in relation to the Rib cage, Pelvis, and Body surface. Identify the following:
   a) Location of Umbilicus (intersection of red lines)
   b) Liver and Stomach (note that both are largely covered by Lower rib cage)
   c) Falciform ligament and the Round ligament of the Liver
   d) Duodenum, Cecum, Appendix
   e) Diaphragm (Thoracic diaphragm)
   f) Kidneys
2. For a more realistic (cadaver) view of the abdominal organs, open **Atlas Anatomy – System – Digestive – Abdominal Viscera.**
3. Also look at the following radiographs (X-ray figures) of the abdominal region taken after swallowing barium. Barium is X-ray opaque and is used to visualize digestive organs.
   Atlas Anatomy – System – Digestive – Barium in Stomach (Ant/Lat)
   Atlas Anatomy – System – Digestive – Barium in Small Bowel
   Atlas Anatomy – System – Digestive – Barium in Large Bowel

LABORATORY 27

The Peritoneum

Section I: ***<u>AnatLab</u>*** – Do Abdomen–Thorax chapters for Peritoneum
Section II: ***ADAM Interactive* – Peritoneum**

Anterior View of Abdomen

1. Open **Dissectible Anatomy** – Start with an Anterior view of the abdomen.
2. Go to depth **layers 158** and **194**. Identify the Parietal layer of the Peritoneum.
3. At depth **layer 195**, identify the following:
   a) Greater Omentum (contains large amounts of fat) – arises from Greater curvature of stomach
   b) Falciform ligament runs from anterior body wall to liver; attaches to liver along line separating Right & Left lobes of Liver
   c) Round ligament of Liver (Runs along inferior margin of the falciform ligament. Remnant of what embryonic structure?)
4. At depth **layers 195** and **199**, identify the Lesser omentum (Spans between what 2 organs?)
5. At depth **layer 200**, identify the Portal triad structures that run through the Lesser omentum, into/out of the Liver:
   a) Common bile duct
   b) Hepatic portal vein
   c) Proper hepatic artery
6. Go to depth **layer 206**. With the stomach removed, identify the following:
   a) Gastrosplenic ligament (*not labeled*; seen as cut edge of Peritoneum that connected Stomach to Spleen). The Splenorenal ligament is not seen in *ADAM*. It would run from the Spleen to the Peritoneum of posterior body wall overlying the left Kidney. Go to depth **layer 211** to identify the relationship of the Spleen to the Kidney (seen here surrounded by Renal fascia).
   b) Transverse mesocolon – supports Transverse colon
7. At depth **layer 208**, identify the Mesentery (supports Jejunum and Ileum portions of Small intestine)
8. Change to depth **layer 210**. Identify the following:
   a) Fascia areas posterior to Ascending and Descending colons – since these portions of the Large intestine are in direct contact with the Posterior abdominal wall (not supported by mesenteric structures), these colon regions are classified as Retroperitoneal structures
   b) Root of Transverse and Sigmoid colons (cut edges of Transverse mesocolon and Sigmoid mesocolon). These are the mesenteric structures that support these colon regions. Thus, these portions of the colon are classified as Peritoneal structures.
   c) Click on central region of Abdominal cavity – labeled as Peritoneum. (What specific layer of the Peritoneum does this represent?)

LABORATORY 28

Blood Vessels and Nerves of the Abdominal Cavity

Section I: ***<u>AnatLab</u>*** – Do Abdomen–Thorax chapters for <u>Vessels and Nerves</u>
Section II: ***<u>ADAM Interactive</u>*** **<u>– Blood Vessels of Abdominal Cavity</u>**

Anterior View of Abdominal Cavity

1. In **Dissectible Anatomy**, open an <u>Anterior view of the male abdominal cavity</u>. Click **Normal Mode** button.
2. Start at depth **layer 201**. Identify the following:
   a) Left gastric artery – to Lesser curvature of Stomach
   b) Proper hepatic artery – to Liver
3. At depth **layer 213**, identify the following:
   a) Branches of the Celiac trunk (*celiac trunk not labeled*):
      1) Left gastric artery
      2) Splenic artery (better seen at depth **layer 216**)
      3) Common hepatic artery – divides into the following:
         (a) Proper hepatic artery – provides hepatic arteries to Right and Left Lobes of liver
         (b) Gastroduodenal artery
   b) Superior mesenteric artery; note how it emerges from behind Pancreas to then pass over the Duodenum (compare at depth **layer 216**). (Organs supplied by this artery?)
   c) Inferior mesenteric artery; emerges from behind Duodenum. (Organs supplied?)
   d) Portal triad structures: Common bile duct, Hepatic portal vein, Proper hepatic artery
4. At depth **layer 217**, identify the following:
   a) Hepatic portal vein (Formed by junction of what 2 veins?)
   b) Superior mesenteric vein
   c) Splenic vein
   d) Inferior mesenteric vein
   e) Left gastric vein
5. At depth **layer 224**, identify the Right and Left gonadal (Testicular in males; Ovarian in female) arteries.
6. Change depth to **layer 232**. Identify the following:
   a) Inferior vena cava – runs along right side of Aorta
   b) Kidneys, Suprarenal (Adrenal) glands, and Renal veins
   c) Hepatic veins (from the Liver)
   d) Formation of Inferior vena cava from union of Right and Left common iliac veins (better seen at depth **layer 241**); Inferior vena cava starts at what vertebral level?
   e) Go back up to depth **layer 222** and identify the following:
      1) Right gonadal vein (Testicular vein in male; Ovarian vein in female); drains directly into Inferior vena cava

2) Left gonadal vein (Testicular or Ovarian vein) – drains into Left renal vein. Left renal vein then crosses mid-line to drain into inferior vena cava.

7. Increase depth to **layer 240** and identify the following:
   a) Abdominal aorta – starts at Aortic hiatus (*not labeled*; opening in Diaphragm for passage of aorta). Note that the Aorta runs down the midline of body (lies directly anterior to Vertebral column).
   b) Unpaired branches of Abdominal aorta:
      1) Celiac trunk (cut)
      2) Superior mesenteric artery (cut)
      3) Inferior mesenteric artery (cut)
   c) Paired branches of Abdominal aorta:
      1) Renal arteries
      2) Gonadal (Testicular in male; Ovarian in female) arteries
      3) Posterior lumbar arteries – one pair per vertebral level to Posterior body wall (better seen at depth **layer 272**)
   d) Termination of Aorta – divides into Right and Left common iliac arteries. At what vertebral level does this division take place?

Section III: *ADAM Interactive* – Nerves of the Abdominal Cavity

Anterior View of Abdominal Cavity

1. Start with an Anterior view of the male abdominal cavity. Go to depth **layer 239** and identify the following:
   a) Branches of Vagus nerve (CN X); enter Abdominal cavity with Esophagus and then pass to anterior side of Abdominal aorta. What specific type of autonomic nerve fibers does the Vagus nerve carry? Where do these fibers synapse?
   b) Thoracic splanchnic nerves (largest is Greater Splanchnic nerve). These start in Thorax region, run through Diaphragm into Abdominal cavity, and then pass to anterior side of Abdominal aorta. (They enter the Abdominal cavity just above the Aortic hiatus.) What specific type of autonomic nerve fibers are carried by splanchnic nerves? Where do they synapse?
2. Go back up to depth **layer 230** and note that the Greater Splanchnic nerve (labeled as Thoracic splanchnic nerve) also send branches into the Suprarenal glands. Which part of the gland is supplied by these nerves? Why?
3. Return to depth **layer 239** and identify the Sympathetic collateral ganglia (located at the base of and named for the major abdominal arteries arising from the aorta):
   a) Celiac ganglion
   b) Superior mesenteric ganglion
   c) Inferior mesenteric ganglion
   d) What type of neurons are located in collateral ganglia? How do the axons from these neurons reach their targets?
4. Go to depth **layer 264** and identify the following:
   a) Sympathetic trunk, with Chain ganglia (specific type of neurons found in these ganglia?)
   b) Splanchnic nerves exiting the thoracic portion of the Sympathetic trunk and running inferiorly.
5. At depth **layer 268**, identify the following:
   a) Quadratus lumborum muscle of posterior abdominal wall
   b) Femoral nerve and Obturator nerve – arise from Lumbar Plexus (spinal nerve levels that form Lumbar plexus?)
   c) Return to depth **layer 266**; note relationships of Psoas major muscle to Femoral nerve (nerve runs lateral to muscle) and Obturator nerve (nerve runs medial to muscle)

LABORATORY 29

Urinary System

Section I: ***AnatLab*** – Abdomen–Thorax chapter, do urinary portion of Urinary and Male Reproductive
Section II: ***ADAM Interactive*** **– Male Urinary System**

Anterior View of Male Abdomen and Pelvis

1. Open **Dissectible Anatomy** – Start with an Anterior view of the Male abdomen and pelvis.
2. At depth **layer 0**, identify the Penis and Scrotum.
3. Go to depth **layer 226** and identify the following:
   a) Bladder – located posterior to Pubic symphysis
   b) Note thick layer of fascia and fat surrounding Kidneys
4. Remove renal fat layers by changing to depth **layer 232**. Identify the following:
   a) Kidneys
   b) Suprarenal (adrenal) glands
   c) Ureters
   d) Hilum (Hilus) of kidney
5. Go to depth **layer 241** and identify the following:
   a) Ureter and Renal pelvis
   b) Major calyx and Minor calyx
   c) Renal cortex with Renal columns
   d) Renal medulla with Renal pyramids

Medial View of Male Abdomen and Pelvis

1. Use **View** button to change to a Medial view of the Male pelvic region. Click **Normal Mode** button and identify the following:
   a) Urinary bladder
   b) Rectum
   c) Rectovesical pouch – depression of Peritoneum between Bladder and Rectum; lowest point of male abdominal cavity
   d) Prostate gland and Prostatic urethra
   e) Sphincter urethra muscle and Membranous urethra
   f) Spongy (penile) urethra

LABORATORY 30

Male Reproductive System; Blood Vessels of the Male Pelvis

Section I: ***AnatLab*** –Abdomen–Thorax chapters, do male reproductive portion of Urinary and Male Reproductive
Section II: ***ADAM Interactive*** **–Male Reproductive System**

Anterior View of Male Abdomen and Pelvis

1. Open **Dissectible Anatomy** – Start with an Anterior view of the Male abdomen and pelvis.
2. At depth **layer 0**, identify the Penis and Scrotum.
3. At depth **layers 24** and **25**, identify the following:
   a) Penis
   b) Spermatic cord (*not labeled*; covered with external spermatic fascia); seen entering the Superficial inguinal ring (*not labeled*) to pass through Abdominal wall via the Inguinal canal
4. At depth **layer 179**, identify the following:
   a) Testis and Epididymis
   b) Ductus deferens and Testicular artery; these run within Spermatic cord through the Superficial inguinal ring (*not seen*), through the Inguinal canal (regions of Ductus and Artery seen running parallel to the Inguinal ligament), and disappear through the Deep inguinal ring (*not labeled*) into the Abdominal cavity
5. Go to depth **layer 226** and identify the Bladder.
6. Go to depth **layer 240** and identify the following:
   a) Bifurcation of Abdominal aorta into Right and Left common iliac arteries (Vertebral level?)
   b) Division of Common iliac artery into Internal iliac and External iliac arteries
   c) Branch of External iliac artery – Inferior epigastric artery (runs up anterior abdominal wall, deep to Rectus abdominis muscle)

Medial View of Male Abdomen and Pelvis

1. Use **View** button to change to a Medial view of the Male pelvic region. Click **Normal Mode** button.
2. Change depth to **layer 42** and identify the following:
   a) Urinary bladder
   b) Rectum
   c) Prostate gland and Prostatic urethra
   d) Corpus spongiosum (with Glans penis region) of Penis and Spongy (penile) urethra
   e) Corpus cavernosum of Penis

f) Testis and Epididymis (in Scrotum – *not labeled*)

3. At depth **layer 49**, identify the following:
   a) Ductus deferens; runs from Testis through Inguinal canal of Abdominal wall with other components of Spermatic cord (Testicular artery and vein; go to depth **layer 56** to see these passing into Deep Inguinal ring). Ductus seen here as it emerges into the Abdominal cavity from the Deep Inguinal ring (*not labeled*) to then pass to the posterior side of the Bladder.
   b) Ampulla and Ejaculatory duct of Ductus deferens; at the Posterior bladder, the Ductus deferens enlarges to form the Ampulla of the Ductus deferens, then narrows to pass through the Prostate gland as the Ejaculatory duct of the Ductus deferens. The Ductus deferens terminates by emptying into the Prostatic urethra.
4. At depth **layer 50**, identify the Seminal Vesicle gland; empties into Ejaculatory duct at posterior side of Prostate gland. Note: Bulbourethral gland not seen in *ADAM*. (Small gland embedded with Sphincter urethra muscle.)
5. Go to depth **layer 65** and identify the following:
   a) Common Iliac artery
   b) External Iliac artery
   c) Internal Iliac artery – branches:
      1) Superior Gluteal artery (exits pelvis through greater sciatic foramen)
      2) Inferior Gluteal artery (exits pelvis through greater sciatic foramen)
      3) Obturator artery (exits pelvis through obturator foramen)
      4) Rectal artery
      5) Vesical artery (to Bladder)

Atlas Anatomy

1. **File – Open – Content – Atlas Anatomy – System – Reproductive – Male Pelvic Organs (Ant)**
2. Identify the following:
   a) Urinary bladder, Ureters, and Urethra
   b) Testis and Epididymis
   c) Ductus deferens
   d) Prostate gland and Seminal vesicle gland (posterior to Bladder)

1. **File – Open – Content – Atlas Anatomy – System – Reproductive – Male Superficial Perineal Space 1**
2. Identify:
   a) Corpus spongiosum
   b) Corpus cavernosum (includes Crus of Penis)
   c) Spongy urethra

3D Anatomy – Male Reproductive System

1. **File – Open – Content – 3D Anatomy – 3D Male Reproductive**
2. Use the pull-down menu to identify the structures of the male reproductive system. (*3D Male Reproductive not available with Student version of ADAM Interactive.*)

LABORATORY 31

Female Reproductive System; Blood Vessels of the Female Pelvis

Section I: *AnatLab* – Do Abdomen–Thorax chapters for Female Reproductive
Section II: *ADAM Interactive* – Female Reproductive System

Anterior View of Female Abdomen and Pelvis

1. Open **Dissectible Anatomy** – Start with an Anterior view of the Female abdomen and pelvis.
2. Start at depth **layer 0**. Identify the Labia majora (majus).
3. Change depth to **layer 219** and identify the following:
   a) Urinary bladder
   b) Ureters (better seen at depth **layers 221** and **233**)
4. At depth **layer 226**, identify the following:
   a) Uterus – Fundus, Body, Cervix
   b) Vagina
   c) Uterine tube – Isthmus, Ampulla, Infundibulum, Fimbria
   d) Ovary
   e) Rectum
5. Go back to depth **layer 223** and identify the Ovarian artery. Note that this arises from the Abdominal aorta.
6. Continue back to depth **layer 207**. Here the Peritoneum is seen covering the pelvic organs. Starting anteriorly, these are the Bladder, Uterus, and Rectum. The recess between the Bladder and Uterus is the Vesicouterine pouch (*not labeled*). The recess between the Uterus and the Rectum is the Rectouterine pouch (*not labeled*). Note how the Peritoneum covering the Uterus passes up the anterior side of Uterus, over its top, and then down the posterior side of the Uterus. This fold of Peritoneum extends laterally off to both sides of the Uterus to also cover the Uterine tubes, Ovaries, and the last portion of the Ovarian artery and vein. The fold then attaches to the lateral walls of the pelvic cavity. This entire fold (spanning between lateral pelvic walls and covering the uterus, Uterine tubes, and Ovaries) is the Broad Ligament of the Uterus (*not labeled*).
7. Change to depth **layer 239** and identify the following:
   a) Bifurcation of Abdominal aorta into Right and Left common iliac arteries (Vertebral level?)
   b) Division of Common iliac artery into Internal iliac and External iliac arteries
   c) Branch of External iliac artery – Inferior epigastric artery (runs up Anterior abdominal wall, deep to Rectus abdominis muscle)

Medial View of Female Pelvis

1. Use **View** button to change to a Medial view of the female pelvis region. Click **Normal Mode** button.
2. Start at depth **layer 0** and identify the following:
   a) Clitoris, Labia majora (majus), and Labia minora (minus)
   b) Urinary bladder, Urethra, and Sphincter urethra muscle
   c) Vagina and Fornix
   d) Uterus – Fundus, Body, Cervix, Cervical canal (*not labeled*), and External Os (*not labeled*). Note angle of joining between Cervix and Vagina. Also note anterior bend between body and Cervix of Uterus.
   e) Rectum
3. Go to depth **layer 46** and identify the following:
   a) Vesicouterine pouch (*not labeled*) – downward fold of Peritoneum between Bladder and Uterus
   b) Rectouterine pouch (*not labeled*) – downward fold of Peritoneum between Rectum and Uterus. This is the lowest (most inferior) point of the female Peritoneal cavity.
4. At depth **layer 48**, identify :
   a) Uterine tube – Ampulla, Infundibulum, Fimbria
5. At depth **layer 69**, identify the following:
   a) Common Iliac artery
   b) External Iliac artery
   c) Internal Iliac artery – branches:
      1) Superior Gluteal artery (exits pelvis through greater sciatic foramen)
      2) Inferior Gluteal artery (exits pelvis through greater sciatic foramen)
      3) Obturator artery (exits pelvis through obturator foramen)
      4) Rectal artery
      5) Vesical artery (to Bladder)
      6) Vaginal artery
      7) Uterine artery

Atlas Anatomy

1. **File – Open – Content – Atlas Anatomy – System – Reproductive – Viscera of Female Pelvis (Sup)**
2. Identify:
   a) Urinary bladder
   b) Uterus – Fundus, Body, Cervix
   c) Uterine tube – Ampulla, Infundibulum, Fimbria
   d) Ovary
   e) Broad ligament of the Uterus
   f) Vesicouterine pouch and Rectouterine pouch

1. **File – Open – Content – Atlas Anatomy – System – Reproductive – Female Pelvic Organs (Ant)**
2. Identify:
   a) Urinary bladder, Ureters, and Urethra
   b) Uterus, Uterine tube, and Ovary

1. **File – Open – Content – Atlas Anatomy – System – Reproductive – Female Superificial Perineal Space 1**
2. Identify the following:
   a) Corpus cavernosum of Clitoris (labeled as Body and Crus of Clitoris)
   b) Vestibular bulb (labeled as Vestibule)
   c) External opening of urethra
   d) Vaginal opening
   e) Anus

1. **File – Open – Content – Atlas Anatomy – System – Reproductive – Contents of Female Pelvis (Sup)**
2. In this cadaver view, identify the following:
   a) Broad ligament of the Uterus
   b) Body of Uterus
   c) Uterine tube
   d) Ovary
   e) Bladder
   f) Rectum
   g) Vesicouterine pouch
   h) Rectouterine pouch

3D Anatomy – Female Reproductive System

1. **File – Open – Content – 3D Anatomy – 3D Female Reproductive**
2. Use the <u>pull-down menu</u> to identify the structures of the female reproductive system. (*3D Female Reproductive not available with Student version of ADAM Interactive.*)

LABORATORY 32

The Thorax

Section I: ***AnatLab*** – Do Abdomen–Thorax chapter for The Thorax
Section II: ***ADAM Interactive*** **– Thorax and Respiratory Tract**

Medial View of Pharynx and Larynx

1. Open **Dissectible Anatomy** – Start with a Medial view of the head and neck region.
2. At depth **layer 0**, identify the following:
   a) Nose
   b) Nasal cavity – Superior, Middle, and Inferior nasal concha; hard palate (*not labeled*) – separates nasal cavity and oral cavity
   c) Nasopharynx – opening of Auditory tube (connects Nasopharynx to Middle ear cavity); Soft palate – elevates during swallowing to close off Nasopharynx from Oropharynx
   d) Oropharynx
   e) Laryngopharynx
   f) Esophagus
   g) Larynx:
      1) Epiglottis
      2) Vocal fold (vocal cord) – formed by underlying Vocal ligament
      3) cut edges of Epiglottic cartilage, Thyroid cartilage, Cricoid cartilage, and Hyoid bone
   h) Trachea

Anterior View of Pharynx and Larynx

1. Use the **View** button to change to an Anterior view of the head and neck region. Click **Normal Mode** button.
2. Go to depth **layer 59**. With the tongue removed, identify:
   a) Soft palate and Uvula (posterior termination of Soft palate)
   b) Posterior wall of Oropharynx
   c) Epiglottis
   d) Hyoid bone
   e) Laryngeal prominence of Thyroid cartilage
3. Change depth to **layer 254**. Identify the following:
   a) Hyoid bone
   b) Thyroid cartilage – connected to Hyoid bone by layer of fascia
   c) Cricoid cartilage
   d) Trachea with Tracheal cartilage rings

4. At depth **layer 255**, identify the Epiglottis and Cricoid cartilage.
5. At depth **layer 256**, identify the Arytenoid cartilages.
6. At depth **layer 257**, identify the following:
   a) Esophagus
   b) Muscles of the Pharyngeal wall (Pharyngeal constrictor muscles; you do not need to know names)
   c) Go back up to depth **layers 256** and **255** and note how the Epiglottis protects the Laryngeal entrance by diverting swallowed food/liquid laterally around the Larynx and then posteriorly into the Esophagus

Anterior View of the Thorax

1. In **Dissectible Anatomy**, open an Anterior view of the chest region.
2. Change depth to **layer 145** and identify the following:
   a) Sternum – Manubrium, Body, Xiphoid process
   b) Ribs and Costal cartilages
   c) External intercostal muscle layer
3. At depth **layer 146**, identify the Internal intercostal muscle layer.
4. At depth **layer 148**, identify the following:
   a) Innermost intercostal muscle layer
   b) Intercostal nerves (Located between what two muscle layers?)
   c) Internal thoracic artery (Branch of what artery?)
      1) Anterior intercostal arteries
      2) Superior epigastric artery (Supplies what muscle?)
      3) Musculophrenic artery (*not seen* – runs along Inferior margin of Rib cage and provides last Anterior intercostal arteries)
   d) Posterior intercostal arteries – from Thoracic aorta
5. Go to depth **layers 158-159** and identify the following:
   a) Diaphragm – attached to Xiphoid process and Anterior margin of Rib cage
   b) Parietal pleura lining inner wall of Thoracic cavity
6. At depth **layer 162**, identify the following:
   a) Right lung – Superior, Middle, and Inferior lobes
   b) Left lung – Superior and Inferior lobes
   c) Parietal pleura
7. At depth **layer 168**, identify the following:
   a) Diaphragm and Central tendon of diaphragm (better seen at depth **layer 176**)
   b) Phrenic nerves
8. Increase depth to **layer 228**. Identify the following in the Diaphragm:
   a) Caval hiatus (*not labeled*) – for passage of the Inferior vena cava (At what vertebral level?)
   b) Esophageal hiatus (*not labeled*) – for passage of the Esophagus (Vertebral level?)
   c) Aortic hiatus (*not labeled*) – for passage of the Aorta (better seen at depth **layer 240**) (Vertebral level?)
   d) Branches of vagus nerve – pass from Thoracic to abdominal cavities by running with Esophagus
9. Go to depth **layer 245**. Identify the following:
   a) Diaphragm; note attachment to Rib 12 (*not labeled*) and bodies of Lumbar vertebra
   b) Cut end of Greater Splanchnic nerves; penetrates Diaphragm near Aorta to enter abdominal cavity

10. At depth **layer 256**, identify the following:
    a) Bifurcation of Trachea into Right and Left mainstem (primary) bronchi – at Plane of Sternal Angle
    b) Lobar (Secondary) bronchi
    c) Segmental (Tertiary) bronchi
    d) Esophagus located posterior to Trachea
    e) Thoracic aorta
11. At depth **layer 262**, identify the Parietal pleura that lines the posterior wall of Pleural cavity.
12. At depth **layer 264**, identify the following:
    a) Azygos vein – receives drainage from entire Thoracic wall; empties into Superior vena cava
    b) Posterior intercostal arteries – one pair per intercostal space; arise from Thoracic aorta
    c) Sympathetic trunk:
        1) Chain (Sympathetic) ganglia
        2) White rami communicantes (Carry what type of nerve fiber? Associated with which spinal nerves?)
        3) Gray rami communicantes (Carry what type of nerve fiber? Associated with which spinal nerves?)
        4) Greater splanchnic nerve (Carries what type of nerve fiber? Destination?)
13. At depth **layer 268**, identify the following:
    a) Intercostal nerve and Posterior intercostal artery and vein in each intercostal space
    b) Innermost intercostal muscle layer
14. At depth **layer 269**, identify the Internal intercostal muscle layer.

Lateral View of the Thorax

1. Use **View** button to change to a <u>Lateral view of the chest region</u>. Click **Normal Mode** button.
2. Change to depth **layer 149** and identify the following:
    a) Diaphragm
    b) Phrenic nerve – passes down along lateral side of Heart, then innervates the Diaphragm
    c) Sympathetic trunk:
        1) Chain (Sympathetic) ganglia
        2) White rami communicantes
        3) Gray rami communicantes
        4) Greater splanchnic nerve
    d) Azygos vein and Superior vena cava
    e) Intercostal nerve and Posterior intercostal artery and vein
3. At depth **layer 204**, identify the following:
    a) Openings through Diaphragm (*not labeled*):
        1) Caval hiatus
        2) Esophageal hiatus
        3) Aortic hiatus
    b) Vagus nerve – crosses anterior to the Subclavian artery, then runs with the Esophagus (posterior to Heart)

3D Anatomy of the Respiratory Tract

1. **File – Open – Content – 3D Anatomy – 3D Lungs**
2. Use the pull-down menu to identify the following:
   a) Parts of Larynx:
      1) Arytenoid cartilage
      2) Cricoid cartilage
      3) Epiglottis
      4) Hyoid bone
      5) Thyroid cartilage
      6) Vocal ligament
   b) Trachea and Tracheal cartilage rings
   c) Lungs:
      1) Right lung – Superior, Middle, and Inferior lobes
      2) Left lung – Superior and Inferior lobes
   d) Right and Left mainstem (main; primary) bronchi

LABORATORY 33

The Heart and Great Vessels

Section I: ***<u>AnatLab</u>*** – Do Abdomen–Thorax chapters for <u>The Heart</u> and <u>Review and Quiz</u>
Section II: ***<u>ADAM Interactive</u>* <u>– Heart and Great Vessels</u>**

Anterior View of Heart and Great Vessels

1. Open **Dissectible Anatomy** – Start with an <u>Anterior view of the chest region</u>.
2. At depth **layer 76**, identify the following:
   a) Right and Left subclavian veins
   b) Right and Left internal jugular veins
   c) Right and Left brachiocephalic veins
3. Increase depth to **layers 169** and **170**. Identify the following:
   a) Pericardial sac – outer wall (seen in **layer 169**) is the fibrous Pericardium layer
   b) Inner wall of Pericardial sac (**layer 170**) is the Parietal layer of the Serous pericardium
   c) Surface of Epicardium (**layer 170**) is the Visceral layer of the Serous pericardium
4. At depth **layer 171**, identify the following:
   a) Right and Left brachiocephalic veins – join together to form Superior vena cava
   b) Superior vena cava (Empties into which chamber of the Heart?)
   c) Ascending aorta and Arch of aorta
   d) Brachiocephalic artery (Divides into what arteries?)
   e) Left common carotid artery and Left subclavian artery
   f) Left vagus nerve (seen crossing over Aortic arch)
   g) Pulmonary trunk
   h) Right and Left pulmonary arteries
   i) Ligamentum arteriosum – adult remnant of embryonic Ductus Arteriosus that connected top of Pulmonary trunk to bottom of Aortic arch, allowing for blood to bypass the embryonic lungs. Seen here at the point where the Vagus nerve disappears between the Pulmonary trunk & Aorta
   j) Heart:
      1) Right atrium and Right auricle
      2) Right and Left ventricles
      3) Base of Heart and Apex of Heart (Apex formed by which Heart chamber?)
      4) Right coronary artery (Seen running in what sulcus?)
      5) Anterior interventricular artery (Runs in what sulcus? Branch of what artery?)
      6) Great cardiac vein (Runs with what artery? Drains into?)
5. Change to depth **layer 173** to identify the Coronary sulcus and Anterior interventricular sulcus.
6. At depth **layer 174**, identify the following inside the Heart:
   a) Right Atrium and Auricle; Pectinate muscle – strands of muscle on inner wall of Auricles

b) Right Ventricle:
   1) Tricuspid valve (Right atrioventricular valve)
      (a) Cusp (*not clearly seen*)
      (b) Chordae tendineae
      (c) Papillary muscle
   2) Pulmonary semilunar valve – with 3 cusps, but no chordae tendineae or papillary muscles

c) Left ventricle:
   1) Trabeculae carneae – strands of muscle on inner walls of Ventricles
   2) Papillary muscle and Chordae tendineae (Associated with what valve?)

d) Note that the muscle of the Left ventricle wall is much thicker than wall of Right ventricle (Why?)

7. Go to depth **layer 175** and identify the cut margins of the following:
   a) Aorta
   b) Superior vena cava
   c) Inferior vena cava
   d) Pulmonary trunk
   e) Pulmonary veins (4)
   f) Posterior wall of the Pericardial sac (Fibrous pericardium lined on its inner surface with Parietal layer of Serous pericardium)
8. At depth **layer 176**, identify the following:
   a) Superior and Inferior vena cava
   b) Ascending, Arch, and Descending Thoracic Aorta
   c) Pulmonary Trunk, and Right & Left pulmonary arteries
   d) Pulmonary veins (2 from each lung) (Drain into what Heart chamber?)

Lateral View of the Heart and Great Vessels

1. Use **View** button to change to a <u>Lateral view of the chest region</u>. Click **Normal Mode** button.
2. Go to depth **layer 210** and identify the following:
   a) Arch of aorta
   b) Brachiocephalic artery
   c) Diaphragm
   d) Cut margins of Superior and Inferior vena cava, Pulmonary arteries, Pulmonary veins
3. Go to depth **layer 218** and identify the following:
   a) Right Atrium and Right Auricle
   b) Right Ventricle
   c) Superior and Inferior vena cava (Drain into which Heart chamber?)
   d) Right pulmonary artery
   e) Left atrium
   f) Right pulmonary veins (2) (Drain into which heart chamber?)
   g) Right coronary artery
   h) Ascending aorta and Arch of aorta
   i) Brachiocephalic artery
4. At depth **layer 220**, identify the following:
   a) Interior of Right atrium
   b) Pectinate muscle of Right auricle
   c) Tricuspid (Right atrioventricular) valve

d) Entrance of Coronary sinus – small opening next to entrance of Inferior vena cava; site at which Coronary sinus drains into Right atrium
e) Fossa ovalis – adult remnant of embryonic Foramen ovale (fetal opening that allowed for passage of blood from Right atrium directly into Left atrium)

Atlas Anatomy

1. **File – Open – Content – Atlas Anatomy – System – Cardiovascular – Heart and Great Vessels (Ant)**
2. Identify the following:
   a) Right Atrium and Right Auricle
   b) Right and Left ventricles
   c) Apex of heart
   d) Right coronary artery
   e) Anterior interventricular artery and Great cardiac vein
   f) Right and Left brachiocephalic veins
   g) Superior and Inferior vena cava
   h) Ascending aorta and Arch of aorta
   i) Pulmonary trunk and Pulmonary arteries
   j) Ligamentum arteriosum
   k) Pulmonary veins
   l) Brachiocephalic artery, Right common carotid artery, Right subclavian artery

1. **File – Open – Content – Atlas Anatomy – System – Cardiovascular – Heart and Great Vessels (Post)**
2. Identify the following:
   a) Left and Right atria
   b) Left and Right ventricles
   c) Posterior interventricular artery (Runs in what sulcus? Branch of what artery?)
   d) Circumflex coronary artery (Runs in what sulcus? Branch of what artery?)
   e) Pulmonary arteries (Pulmonary trunk not seen in this view)
   f) Pulmonary veins (2 from each lung)
   g) Arch of aorta, Brachiocephalic artery, Right common carotid artery, Right subclavian artery
   h) Superior and Inferior vena cava
   i) Azygos vein – drains into Superior vena cava

1. **File – Open – Content – Atlas Anatomy – System – Cardiovascular – Coronary Arteries (Sup)**
2. Identify the following:
   a) Right coronary artery
   b) Posterior interventricular artery
   c) Left coronary artery
   d) Anterior interventricular artery
   e) Circumflex coronary artery
   f) Left atrioventricular (bicuspid, mitral) valve and valve cusps
   g) Right atrioventricular (tricuspid) valve and valve cusps
   h) Aortic semilunar valve and valve cusps
   i) Pulmonary semilunar valve and valve cusps

1. **File – Open – Content – Atlas Anatomy – System – Cardiovascular – Dissection of Heart Chambers**
2. In this cadaver view, identify the following:
   a) Right atrium with the Fossa ovalis
   b) Superior and Inferior vena cava
   c) Right ventricle with Trabeculae carnea
   d) Interventricular septum
   e) Left atrium
   f) Left atrioventricular (bicuspid, mitral) valve
   g) Chordae tendinea
   h) Papillary muscle
   i) Left ventricle (note thicker muscular wall compared to wall of right ventricle)
   j) Apex of Heart

3D Anatomy

1. **File – Open – Content – 3D Anatomy – 3D Heart**
2. Use the pull-down menu to identify structures of the Heart.